Managing Biogas Plants

Driven by the increasing focus on sustainable energy production and recent advancements in anaerobic digestion technology, the Second Edition of *Managing Biogas Plants: A Practical Guide*, provides insights into optimizing biogas plant operations. This updated edition explores the latest developments in biochemical methane potential (BMP) testing, automatic measurement systems, and biological process management for industrial biogas plants. Drawing from extensive field experience, the book bridges theoretical concepts with hands-on applications.

Features:

- Includes updated content on state-of-the-art BMP testing methods and their practical applications in energy production.
- Incorporates the latest technology in automatic methane potential measurement, including third-generation instruments with improved accuracy and capabilities.
- Provides critical analysis of current industry standards and presents enhanced sections on error propagation analysis and uncertainty margins in BMP testing.
- Equips professionals with proper methodologies for expressing BMP test results for business planning and decision-making.
- Updates practical applications with current case studies reflecting real-world implementation.

Complete with updated figures, practical examples, and troubleshooting guidance, this book is aimed at biogas plant managers, researchers, and industry professionals.

Green Chemistry and Chemical Engineering

Series Editor
Dominic C.Y. Foo
University of Nottingham, Malaysia

Water for Energy and Fuel Production
Yatish T. Shah

Managing Biogas Plants: A Practical Guide
Mario Alejandro Rosato

The Water-Food-Energy Nexus: Processes, Technologies, and Challenges
I. M. Mujtaba, R. Srinivasan, and N. O. Elbashir

Hemicelluloses and Lignin in Biorefineries
Jean-Luc Wertz, Magali Deleu, Séverine Coppée, and Aurore Richel

Materials in Biology and Medicine
Sunggyu Lee and David Henthorn

Resource Recovery to Approach Zero Municipal Waste
Mohammad J. Taherzadeh and Tobias Richards

Hydrogen Safety
Fotis Rigas and Paul Amyotte

Nuclear Hydrogen Production Handbook
Xing L. Yan and Ryutaro Hino

Water Management: Social and Technological Perspectives
Iqbal Mohammed Mujtaba, Thokozani Majozi and Mutiu Kolade Amosa

Efficiency and Sustainability in the Energy and Chemical Industries: Scientific Principles and Case Studies, Third Edition
Krishnan Sankaranarayanan

Process Integration for Resource Conservation, Second Edition
Dominic C.Y. Foo

Industrial Decarbonization: Materials, Methods, and Developments
Qingxu "Bill" Jin, Ida Chen, Fan Shi and Dushyant Shekhawat

Valorisation of Food Waste: Process and Product Design
Chien Hwa Chong

Managing Biogas Plants: A Practical Guide, Second Edition
Mario Alejandro Rosato

For more information about this series, please visit: www.routledge.com/Green-Chemistry-and-Chemical-Engineering/book-series/CRCGRECHECHE

Managing Biogas Plants

A Practical Guide

Second Edition

Mario Alejandro Rosato

CRC Press
Taylor & Francis Group
Boca Raton London New York

CRC Press is an imprint of the
Taylor & Francis Group, an **informa** business

Designed cover image: Shutterstock

Second edition published 2026
by CRC Press
2385 NW Executive Center Drive, Suite 320, Boca Raton FL 33431

and by CRC Press
4 Park Square, Milton Park, Abingdon, Oxon, OX14 4RN

CRC Press is an imprint of Taylor & Francis Group, LLC

First edition published by CRC Press 2017

ISBN: 978-1-04-114797-8 (hbk)
ISBN: 978-1-04-114798-5 (pbk)
ISBN: 978-1-00-367620-1 (ebk)

DOI: 10.1201/9781003676201

Typeset in Palatino
by Newgen Publishing UK

Contents

Preface

Mario Alejandro Rosato

In recent years, the anaerobic digestion industry has gained momentum due to new geopolitical tensions and stringent decarbonization policies. While biogas plant design and construction is a mature technology, its deployment at global scale poses some challenges like dealing with "exotic" feedstock unknown to the technology providers, NIMBY (not in my back yard) opposition to new projects, lack of harmonized protocols on measuring methane leaks (the so-called fugitive emissions), and hard-to-die reliance on useless techniques like monitoring the pH or the FOS/TAC (aka VFA/TA) of the digestate. At the laboratory scale, the technological progress has increased the accuracy and precision of both instruments and test protocols. Despite the 2018 edition being included in Book Authority's list of the 100 best books on industrial chemistry in 2022 and remaining on it until 2025, the text has become obsolete. This second edition is meant to close the gap and to update the anaerobic digestion professionals—both academic and industrial ones—to the latest standards, protocols, and laboratory measurement technologies.

For purely didactic reasons, some products and commercial brands will be quoted throughout this book. The inclusion of said references was a consequence of the simple fact that either the author had direct experience in the use of the said products or that publishable material or information was available (including authorization to publish pictures, tables, etc.). Consequently, the said references do not constitute a recommendation for their purchase or an opinion as to their quality. All brands quoted are marked with ® and belong exclusively to their owners.

General Scope of This Book

There are three mutually independent but complementary strategies to optimize biogas plants:

1 *Optimum design of the plant*: The design of an existing biogas plant is not always optimal. There are many reasons for this, but the most common are the following:

   a The standardization of the components, intending to obtain economies of scale as a function of the plant's size that the manufacturer considers to be optimal for his market conditions.

- b Changes in norms and regulations after the plant's construction.
- c Unforeseen variations in feedstock prices.
- d Sometimes, more often than one could expect, the ignorance and superficiality of the designer and reliance on BMP (biochemical methane potential) data from the literature.

2 *Increasing the net yield of methane per mass unit of feedstock.* Sometimes, a biogas plant may be specially designed to digest a specific substrate. Even in such a condition, it is always possible to extract more methane from the feedstock with the help of particular techniques, such as thermal or chemical pretreatment, milling, or other methods of reducing particle size, sonication, and the addition of enzymes. On the other hand, improving the digestion efficiency usually comes at the expense of consuming additional energy and/or chemicals, so the EROI (energy return on investment) may not be profitable, in some cases, even negative. To complicate things, literature on calculating the EROI of anaerobic digestion plants tends to be confusing and often incorrect (e.g., computing the energy to heat the digester as an "investment" in plants where such energy is an "internal byproduct" because it is recovered from the motor's residual heat).

3 *Maximizing the methane yield per unit volume of the digester.* Another way to improve the performance of the biogas plant is to vary the parameters of the digestion process, such as its temperature, stirring intensity, and dilution of the feedstock. Again, this usually comes at the expense of consuming more energy, so a thorough calculation of the EROI is needed.

Concerning the strategies outlined above, this book has three goals:

1. Providing the reader with a clear idea about what parameters are worth measuring.
2. Defining what instruments and techniques are the most suitable for such measurements.
3. Providing some guidance for the correct planning of the laboratory tests, based on the available instruments and human resources, to minimize the uncertainty of the results.

The author kept the theory to a minimum, prioritizing practical study cases showing how to manage the anaerobic digestion process as reliably as possible. The method, developed by the author and described in this manual, requires a small investment in suitable laboratory equipment. Such investment is marginal in comparison to the total cost of the biogas plant itself or the operational cost of laboratory tests performed by external providers.

Despite the primarily practical approach of this volume, readers should thoroughly study the theoretical sections in each chapter, as they are essential for understanding and deconstructing many of the myths and legends that are common—and still alive—seven years after the first edition's publication. Most of these arise from a partial or incomplete interpretation of the data available in the scientific literature. In contrast, others are sometimes the product of marketing pressure from companies that exaggerate the performance of their products. By focusing on the day-to-day, practical problems of biogas plant management, this book is a valuable instrument for those readers who do not necessarily have a professional background in biological or chemical laboratory techniques. The reader will learn, step by step, how to perform routine tests using a minimal kit of instruments and how to apply the practical notions contained in this book to a successful professional activity.

Final Caveat

As of 2025, people tend to rely on information obtained from several artificial intelligence (AI) apps widely available to the public. The author's experience with such tools is negative, as they tend to provide excessively superficial, often incorrect, information retrieved from blogs, chats, and even scientific literature, which is far from perfect. No AI has been employed to write this book. The content derives mainly from tests conducted by the author himself, and from a few selected literature sources reputed as reliable.

The author gained his experience by operating in the Italian biogas market, where most of the biogas plants run on agricultural feedstock. The Italian biogas market (the second largest in Europe by installed power) is strongly influenced by the German biogas industry and European policies. Agricultural biogas plants in Germany and Italy primarily aim at energy generation, as this is strongly subsidized in these countries. In such contexts, optimization means "getting the maximum amount of energy per ton of feedstock." Plants operating in other countries, or those involved in wastewater or urban garbage treatment, may not share the same goal. For instance, the optimization of an urban garbage biogas plant, which generates income from municipal tipping fees, may sometimes require sacrificing part of its energy productivity in order to achieve a quicker digestion process or safer waste sanitation.

According to the IEA's (International Energy Agency) Global Report on Biogases in 2024, the European Union has produced nearly half of the biogas in the world, followed by China. Nevertheless, this does not mean that European technology, tradition, and policies will be suitable in other geographic regions. For instance, while the European Commission has historically subsidized industrial-scale biogas plants, thereby ignoring and indirectly hindering the deployment of household digesters in Europe, China and India have focused on household-sized digesters due to their lower cost and higher efficacy in replacing fossil fuels. Many multinational companies promote their products in Europe by sponsoring research at a few European universities, as the prestige of these institutions serves as a form of endorsement for the quality of their products. Hence, quite often the European literature on biogas is biased either by marketing or by ideology, and, in some cases, it is self-referenced too. The author's critical approach to techniques, recipes, and information from the biogas literature, as well as how to debunk false or biased information, is condensed in his latest volume, "Optimising biogas plants – Squeezing profit out of your digester". It is worth repeating here the same advice: test in your laboratory to see if other people's findings apply to your case and verify the accuracy and precision of your tests. Never test new methods, feedstocks, or additives directly in the biogas plant.

Each anaerobic digester is a unique ecosystem of microorganisms, including bacteria, protozoans, archaea, and yeasts. The equilibrium between these is variable in time and hard, if not impossible, to predict. It cannot be taken for granted that a result published by a research group, or in the many examples contained in this book, will work or bring a benefit to the reader's digester in a given moment. The following list summarizes the author's advice:

1 Do not forget that every measure, even those taken by famous university laboratories, is subject to errors. The concept of "measurement error" is still taboo for most researchers, and even the German norm VDI 4630/2016 contains a conceptual confusion between accuracy and precision. Hence, always be cautious about adopting biogas plant management methods based on tables.

2 Plan and conduct tests to know exactly how the biological process is running in your anaerobic digester, at this precise moment. Tests you may have performed in the past do not necessarily reflect the current state of the biological ecosystem or the quality of the current feedstock lot. Keeping trackable records of all your tests will allow you to identify trends that may serve as an early warning, enabling you to correct potential process disruptions before it's too late.

3 Measure the yield of your locally available feedstock, lot by lot, using inoculum taken from your biogas plant and including a positive control (acetate or cellulose) to assess the inoculum's methanogenic activity and hence the validity of your test. If you purchase feedstock for your biogas plant, knowing its exact methane yield will enable you to negotiate its price more effectively.

Acknowledgments

To my wife Giovanna, without whose patience and help, this book would never have come to press.

To my students and customers, whose questions and comments have contributed to making this text more complete and easier to read.

To Professor Jing Liu and all the staff of BPC Instruments AB (formerly Bioprocess Control AB), for the photos and advice provided.

To Professor Raman Saravanane for the valuable advice and bibliographic material provided.

To Dr. Flávio Ascenco and the staff of the World Biogas Association for their contribution to disseminating the knowledge of anaerobic digestion.

Portogruaro (Italy)
July 2025

About the Author

Mario Alejandro Rosato is an electrical, electronic, and environmental engineer, as well as a scientific journalist. He built his first home-sized digester in Argentina when he was 16 years old. The son of a university professor and researcher, since his childhood, he learnt to employ sophisticated instruments in his favorite playground—his father's lab. At the age of 25, he obtained a scholarship to specialize in renewable energies in Italy. At 28, he chose to leave the academic research world and devoted himself to developing industrial solutions. In 1990, he settled in Italy, and in 2000, he moved to Spain. In 2004, he became the scientific director of Bioenergía Aragonesa SL. In 2006, he patented the AFADS system, a multitrophic bioreactor for wastewater treatment. In 2009, he founded Sustainable Technologies SL in Barcelona (Spain) and moved its headquarters to Portogruaro (Italy) in 2019. Since 2010, he has provided technical training to biogas researchers and professionals in several private institutes in Italy and Spain, expanding this activity to the rest of the world in collaboration with the World Biogas Association since 2022. Mario A. Rosato has been running a laboratory for applied research and routine anaerobic digestion tests in Italy since 2011. Since 2014, he has been responsible for the column on bioenergy for Agro Notizie, a specialized e-zine catering to Italian agronomists and agricultural entrepreneurs. In 2015, he became a member of the technical commission responsible for drafting the Italian norm on the biochemical methane potential (BMP) test protocol, with his contributions focusing on error propagation analysis and improving both accuracy and precision. In 2022, the first edition of this volume was included in Book Authority's lists of the best 100 books and 20 e-books on chemical engineering.

1

Relevant Aspects for Optimizing the AD Process

1.1 What Is Anaerobic Digestion?

1.1.1 Theoretical Notions

Anaerobic digestion (AD) is a biological process that degrades organic matter in the absence of oxygen, or the absence of air. Because air consists of approximately 21% oxygen, the anaerobic process is sometimes referred to as an anoxic process.

Several types of anaerobic fermentative processes exist in nature:

- Alcoholic fermentation, e.g., during the production of wine or beer.
- Lactic fermentation, e.g., during the production of *sauerkraut* (*choucroute*), corn silage, and soy sauce.
- Dark fermentation. This is a specific type of anaerobic fermentation that produces biohydrogen and requires energy to be supplied in the form of heat.
- Photofermentation. Certain bacteria and cyanobacteria produce biohydrogen, but only when energy is supplied to them in the form of light.
- Butyric fermentation. This is a specific type of dark fermentation that produces hydrogen, butanol, and acetone simultaneously.
- Propionic fermentation. This is the production of propionic acid by *Propionibacterium acidipropionici*, grown under strictly anaerobic conditions using lactose as the carbon source, at a pH of 6.5 and 30°C.
- Methanogenic fermentation. This process is discussed in detail in this book; however, some of the techniques presented here are also suitable for optimizing other fermentative processes.

DOI: 10.1201/9781003676201-1

The organisms directly involved in producing methane are the archaea, which are very similar to bacteria but belong to a different kingdom and are likely the ancestors of bacteria.

The kingdom Archaea encompasses two subkingdoms: the *Crenarchaeota* and the *Euryarchaeota* (Figure 1.1).

The *Crenarchaeota* are known as extremophile organisms because they typically thrive in environments such as submarine volcanoes, sulfur lakes, saline lakes, or even deep rocks, where the temperature, pH, pressure, or salinity reaches values that would be lethal for other life forms. The use of these organisms in industrial applications began in 2014, as researchers discovered some promising species capable of producing ethanol directly from cellulose or repairing their DNA when damaged by very intense radiation. As of July 2025, despite the volume of papers published, the author is not aware of any industrial facility that employs extremophiles.

The *Euryarchaeota,* in turn, encompass all the species of methanogenic archaea known until now. The biodiversity within this subkingdom is remarkable, with some species thriving in the cold seabed of the Arctic, others preferring temperatures near 100°C, and others obtaining their nourishment from hydrogen, which is toxic to other species that feed on acetic acid.

The AD process is a sequential combination of fermentative processes, with methanogenic fermentation being the final stage. As of 2025, our knowledge of the AD processes can be sketched as shown in Figure 1.2.

The organic matter is composed of carbohydrates (compounds of carbon, hydrogen, and oxygen, namely, simple sugars and polysaccharides), proteins

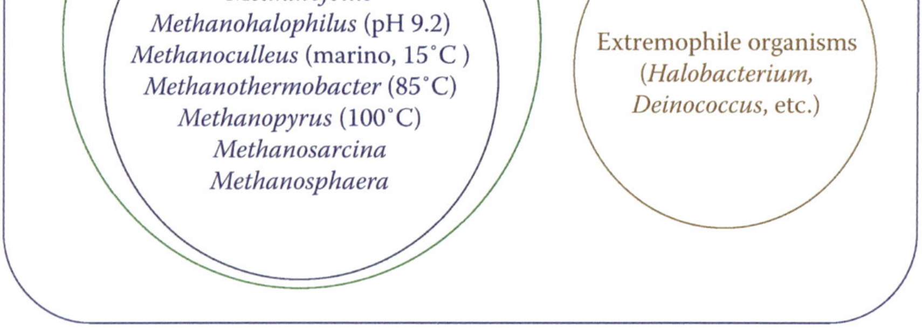

FIGURE 1.1
Situation of the methanogenic organisms within the kingdom Archaea.

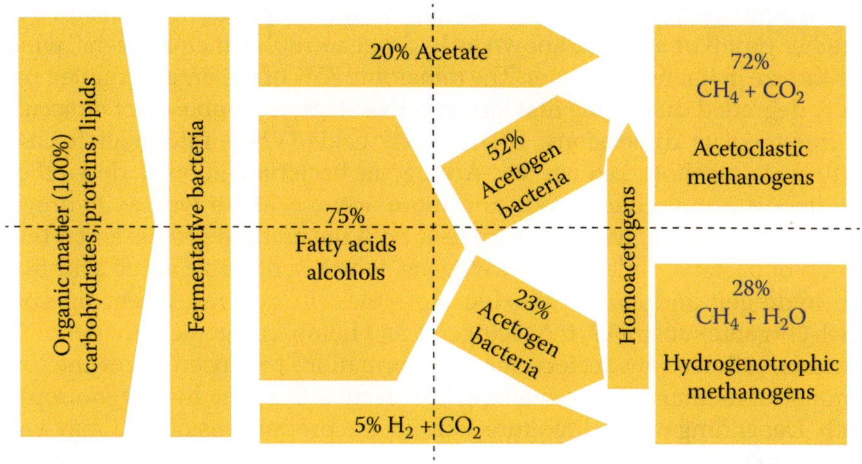

FIGURE 1.2
The phases of the AD process, according to Varnero Moreno (2011). (Varnero Moreno (2011; graphic re-elaboration by the author.)

and amino acids (compounds of carbon, hydrogen, oxygen, and nitrogen, sometimes also sulfur), and lipids (long chains of carbon and hydrogen, with small proportions of oxygen). Furthermore, all organic substances contain a mineral fraction, called ash. The biological decomposition of carbohydrates, proteins, and lipids cannot be taken for granted: some substances, like lignin (a very complex carbohydrate), are indigestible by anaerobic organisms. Other substances like keratin (the protein that composes hair and feathers) are tough to degrade anaerobically. Generally, the more complex the substance, the higher the number of microorganisms involved in its degradation, and hence the longer the time for its complete AD. It is known that some pesticides, which are not degradable through aerobic processes and are very difficult to degrade chemically, can be decomposed by anaerobic bacteria. However, this process takes a considerable amount of time.

Observing Figure 1.2, we notice that the fermentative bacteria and other microorganisms (protozoa, yeasts, and a specific group of fungi called Neocallimastigomycetes, primarily found in the digestive tracts of herbivores, where they play a crucial role in breaking down plant material) are responsible for the first demolition step of the organic molecules. Assuming the total amount of organic matter under fermentation is 100%, 20% of it becomes acetic acid and reacts with the salts present in the fermentation medium, forming acetate that, in turn, feeds a group of methanogens, called acetoclastic methanogens. These latter belong to the kingdom Archaea and thrive by converting the acetate into methane (CH_4) and carbon dioxide (CO_2). Together with the acetic acid, the first fermentative stage transforms 5% of the initial organic matter into free hydrogen (H_2) and CO_2. Both gases, dissolved in the aqueous phase of the fermentation broth, serve as a food source for another group of archaea, known as hydrogenotrophic methanogens, which synthesize methane and water. The remaining 75% of the organic matter, partially degraded during the first fermentative stage, is composed of molecules of intermediate dimensions: volatile fatty acids (VFA), intermediate-chain fatty acids, alcohols, ketones, etc. Acetogenic bacteria will break down these intermediate molecules, producing more acetic acid, as well as hydrogen and carbon dioxide, which in turn feed both the archaea groups and a third group of bacteria, called homoacetogens, which synthesize acetic acid from the hydrogen and carbon dioxide molecules. In general, when digesting "rich" organic substrates, the acetogenic and homoacetogenic activities of the bacteria, followed by acetoclastic transformation, produce 72% of the total amount of methane. The remaining 28% results from the hydrogenotrophic path. Depending on the literature sources, the proportions quoted may vary slightly from author to author, with some of them estimating 70% of the total amount of methane produced through the first pathway and 30% through the second.

The popular wisdom says that "Man does not live by bread alone", and this applies to bacteria too. In the same way that superior animals need a balanced diet, the microbes in a biogas plant require particular attention to the preparation of their "menu", too. Fortunately, the ideal diet for a species or bacterial group is a bit easier to formulate than that for a cow or a person. If we consider the digester as an ecosystem of microorganisms, the ideal diet should have a ratio between carbon and nitrogen (C/N) as close as possible to 30. Since phosphorus is also an essential element for life, the microorganisms' diet must take it into account, resulting in optimum growth when the ratio of C/N/P is equal to 150/5/1. The reason why it is practically impossible to formulate the diet of the biogas plant based only on the C/N/P proportions is simply that the said elements must not only be present in the feedstock, but must also be in a form that is readily assimilable for the microorganisms. Like people or animals, microorganisms are also healthy when they feed on light substances and, conversely, they can even die of indigestion if their diet is too "heavy". The *emoticons* in Table 1.1 help us to remember this important concept.

From the stoichiometric point of view, the reactions leading to the anaerobic degradation of pure substances are provided in Table 1.2.

The value δG_0 represents the total energy consumed (if negative) or produced (if positive) by the reaction. We can observe from Table 1.2 that all chemical reactions of the anaerobic degradation known to date show a negative sign. This means that the reaction needs heat to take place, and hence it is called an endothermic reaction. Depending on the temperature of the heat supply, the AD process can be classified as:

- Psychrophilic, for T between 10°C and 20°C ± 2°C
- Mesophilic, for T between 20°C and 40°C ± 1°C

TABLE 1.1

Digestibility of the Organic Substrates, from Best to Worst

Light carbohydrates (sugars, starch)	☺
Heavy carbohydrates (polysaccharides)	😐
Proteins	😐
Fats, soaps	☹

TABLE 1.2

Stoichiometry of the AD

No.	Reaction	δG_o [kJ/mol CH_4]
1	$4CH_3OH \rightarrow 3CH_4 + CO_2 + 2H_2O$	–106
2	$CH_3OH + H_2 \rightarrow CH_4 + H_2O$	–112.5
3	$4CH_3NH_2 + 2H_2O \rightarrow 3\,CH_4 + CO_2 + 4NH_3$	–76.7
4	$2(CH_3)_2NH + 2H_2O \rightarrow 3CH_4 + CO_2 + 2NH_3$	–74.8
5	$4(CH_3)_3N + 6H_2O \rightarrow 9CH_4 + 3CO_2 + 4NH_3$	–75.8
6	$2(CH_3)_2S + 2H_2O \rightarrow 3CH_4 + CO_2 + 2H_2S$	–52.1
7	$4(CH_3)SH + 2H_2O \rightarrow 3CH_4 + CO_2 + 4H_2S$	–51
8	$(CH_3)SH + H_2 \rightarrow CH_4 + H_2S$	–69.3
9	$4H_2 + CO_2 \rightarrow CH_4 + 2H_2O$	–130.4
10	$CH_3COO^- + H^+ \rightarrow CH_4 + CO_2$	–36
11	$4CO + 2H_2O \rightarrow CH_4 + 3CO_2$	–211

- Thermophilic, for T between 40°C and 60°C ± 0.5°C
- Extremophilic (over 60°C, when other kinds of archaea and bacteria produce mostly H_2, methanogens are inhibited)

N.B.: Some authors define the range 40°C–80°C as thermophilic and do not define an extremophilic range. This is a matter of convention and the age of the publication, given that extremophilic processes are a relatively new field of research.

According to a study by Gerardi (2003), the optimum temperature ranges vary by genus of archaea, as summarized in Table 1.3.

Hence, according to the said research, the maximum biodiversity of archaea species corresponds to a very narrow temperature range between 37°C and 40°C. The maximum biodiversity of the methanogenic ecosystem leads to a stable operation of the AD plant (biogas plant), less prone to "jumps" of the biogas production and more resilient to disturbing events (e.g., sudden changes in the composition of the digester's diet, excursions of the pH, and accidental input of inhibiting substances). The former concepts seem to justify the choice of many biogas plant constructors, who promote mesophilic processes as the most reliable in the market.

TABLE 1.3

Ideal Temperature Ranges for Each Genus of Archaea

Genus	20°C	25°C	30°C	35°C	40°C	45°C	50°C	55°C
Methanobacterium				**	*****	*****		
Methanobrevibacter				**	*****			
Methanosphaera				*****	*****			
Methanolobus				*****	*****			
Methanococcus				*****	*****			
Methanosarcina			*****	*****	*****			
Methanocorpusculum			*****	*****	*****			
Methanoculleus				*****	*****			
Methanogenium	*****	*****	*****	*****	*****			
Methanoplanus			*****	*****	*****			
Methanospirillum				*****	*****			
Methanococcoides			*****	*****				
Methanolobus				*****	*****			
Methanohalophilus				*****	*****	*****		
Methanohalobium							*****	*****
Methanosarcina							*****	*****

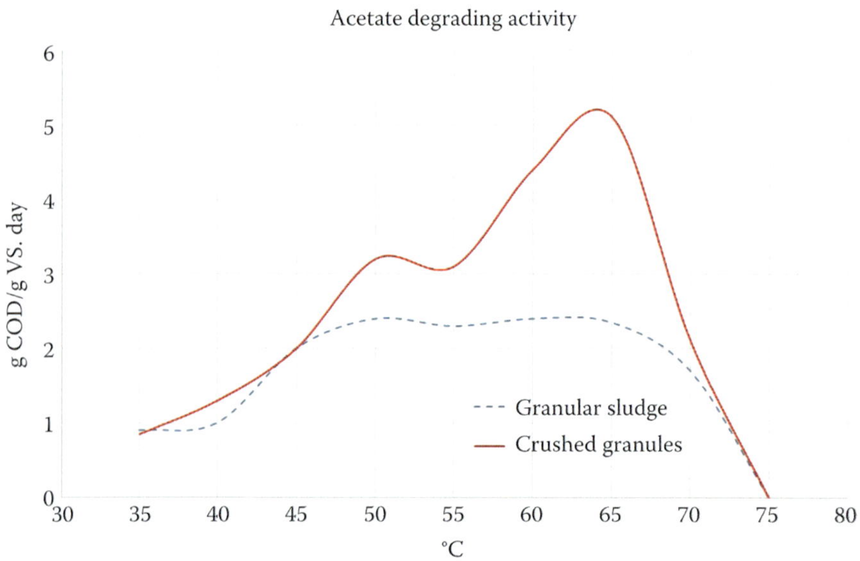

FIGURE 1.3
Anaerobic degradation speed of acetate as a function of temperature, measured on two different types of inocula. (Van Lier, J., et al., *Water Research*, 30(1), 199–207, 1996; graphics by the author.)

According to another study conducted by van Lier et al., the maximum anaerobic degradation rate of acetate corresponds to an incubation temperature of approximately 60°C–65°C, and methanogenic activity varies with the fermentation temperature, as shown in Figure 1.3.

The results of both studies seem to confirm the choice of some biogas plant constructors, who claim that their thermophilic plants, operating at approximately 50°C, are the most efficient (due to their shorter hydraulic and solid retention times, resulting in a more compact design). We will analyze these two (mesophilic vs. thermophilic) in depth in the forthcoming chapters. At this stage, we can anticipate that both are partially correct, because in practice, nobody can demonstrate that one is better than the other in all cases. Remember that the AD is a sequential process; hence, the overall digestion speed is limited by the slowest stage. In general, fermentative bacteria are mesophilic, whereas the archaea tend to be thermophilic, but not all of them. Hence, a digester that optimizes the living conditions of one microbial group will hinder the growth of the other, and vice versa.

1.1.2 Practical Implications

Based on the previous explanations, we can draw the following fundamental practical conclusions.

1.1.2.1 Complexity of the System

The AD is a highly complex process in which more than 1,000 species of microorganisms must coexist, each with different vital needs. For instance, the fermentative *bacteria* prefer acid (low) pH, while the archaea grow well if the pH is kept as close as possible to neutrality. In other words, managing a biogas plant means becoming a "bacteria farmer". Such a job has some similarities to that of the pig or the chicken farmer. It is necessary to feed the bacteria in the digester with the correct diet and maintain the hygienic conditions of their ecosystem. Never forget that microorganisms are as sensitive as superior animals to the excess and deficit of feedstock, to the presence of pollutants (in general, toxic or at least inhibitors of the metabolic activity), and to extreme temperature. The most significant difference between farming bacteria and farming animals is that the "fodder" for the digester can generally be a single substance that will be digested by an ecosystem composed of over a thousand different species, some of which may be more or less efficient in degrading the said feedstock. A biogas plant is also an ecosystem that exhibits higher complexity and biodiversity than a natural park. Hence, the correct management of a biogas plant is not just "throwing waste into the digester". In contrast, it is necessary to calculate in each circumstance, as precisely as possible, which substrates and in which proportion must be fed into the digester, to obtain the desired energy production. Furthermore, it is necessary to prepare each substrate in the best way before feeding it to the digester, to enhance its total conversion into biogas, or at least its highest rate of conversion, since most substrates are not fully degradable (Figure 1.4).

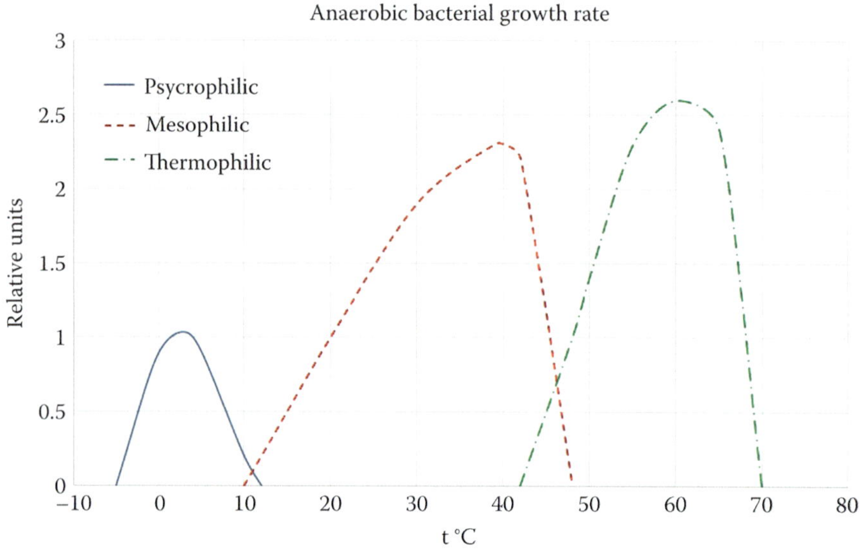

FIGURE 1.4
Relative reproductive activity of the different species of archaea, according to Madigan and Martinko (2006). (Madigan and Martinko (2006), republished by Schnürer and Åsa (2010); graphics by the author.)

1.1.2.2 Sequential Process

The AD process is a sequence of precise biochemical reactions. It is common sense that "a chain cannot be stronger than the weakest of its rings", so a single perturbation (e.g., feedstock containing antibiotics, sudden variations of temperature or pH, etc.), introduced at any trophic level, may lead the anaerobic process to collapse.

1.1.2.3 Multiparametric Process

For the same reasons already explained, there is *no straightforward* way to monitor the AD process, nor is it possible to conduct a census of the microorganisms to determine which are active and which are not. Hence, it is necessary to perform *indirect tests* to understand the population consistency and the biological activity of each group of microorganisms composing the "ecosystem" in the digester.

1.1.2.4 Two Degradation Paths of the Organic Matter

In many books and articles, the AD process is still represented as exclusively limited to the acetogenic–homoacetogenic–acetoclastic path, ignoring the hydrogenotrophic path, whose 28% contribution to the total theoretical methanogenic production should not be overlooked. This conception, at least 30 years old, persists in the design of many biogas plants (single-tank or two identical tanks in parallel), which are then born suboptimal. Being almost impossible to modify a biogas plant after it is already built, it becomes critical to determine, through small-scale tests, the best operational conditions for each reactor, settling for a compromise in the choice of the process parameters (T, pH, stirring, etc.), as will be discussed in the forthcoming chapters.

1.1.2.5 Influence of Temperature

Temperature is a crucial factor, probably the one that has generated the most significant number of "rural legends" among designers, builders, and managers of biogas plants. In practice, the ideal operation temperature of the biogas plant will depend on the composition of the bacterial ecosystem, considered as a whole. Consequently, it is too restrictive to determine the operation temperature solely based on the acetate degradation test, as shown in Figure 1.3, because this test overlooks 28% of the total methane production from the hydrogenotrophic pathway. Nor is the biodiversity of the sole archaea the best criterion to decide the working temperature of a digester. Several other microorganisms must degrade the biomass before the archaea can produce methane, and the ideal temperature for said microorganisms may differ from that of the archaea. From a practical point of view, the usual inoculum to start the digester's operation is cow dung, or a mixture of different farm dung and slurry. All farm animals are hot-blooded, so the microbial flora in their guts has evolved over millions of years to thrive in an environment with a constant temperature in the range of 36–38°C. Hence, operating at lower or higher temperatures will always disrupt the microbial ecosystem's equilibrium. Once again, each biogas plant is a unique case, requiring practical tests to determine the ideal operating temperature for that specific combination of bacterial consortium and feedstock.

1.1.2.6 Need for a Balanced Diet for the Microorganisms

The ideal C/N ratio is 30:1; nevertheless, not all the carbon or the nitrogen present in the feedstock is necessarily digestible. Consequently, it is possible to obtain good results by the AD of substrates having C/N ratios ranging from 10:1 to 90:1. Furthermore, potassium, calcium, magnesium, and other elements play a role, within certain limits, in maintaining a healthy bacterial ecosystem. However, there is considerable marketing hype surrounding their importance, as some companies and biogas plant builders have built a business on selling mineral additives or "boosters". When the feedstock for a biogas plant consists of a mixture of different biomasses, it is advisable to test different proportions to determine the optimum combination. It is not advisable to rely only on the elementary chemical composition of each biomass, as measured by animal feed laboratories or by Near InfraRed Spectrometry (NIRS) apparatus, nor on data

TABLE 1.4

Approximate Contents of C, N, and C/N Ratios of Some Common Biomasses (% WM = Percent of Wet Mass)

Biomass	N (% WM)	C (% WM)	C/N
Corn (whole plant)	1.2	68	56.6
Cabbage	3.6	45	12.5
Grass (hay)	4	48	12
Alfalfa	2.5	40	16
Hay (leguminous)	1.6	40	25
Cow dung	1.6–1.8	30–40	17–25
Sheep dung	3.8	49–76	13–20
Stable dung (with straw)	0.8	22	27.4
Horse dung	2.3	57.5	25
Swine dung	2.8–3.8	23–38	6.2–13.7
Swine manure	0.4	4	10
Straw (oat)	0.5	40	80
Straw (wheat)	0.5	50	100
Straw (barley)	1	48	48
Hen dung (layers)	3.7–6.3	31–35	5–9.6
Tomatoes	3.3	41	12.5
Kitchen waste	1.9	54	28.6
Corn stalks	1.4	44	31
Clover	3	39	13

from the scientific literature. The reason is that the anaerobic degradability of a given biomass also depends on other factors, like the feedstock's age and conservation method, eventual pretreatments, etc. A general idea on the C/N ratio of the different biomasses composing a mixture is useful anyway, at least to calculate the proportions that are nearer to the optimum, to minimize the amount of biological tests to be performed before deciding the digester's diet. Table 1.4 helps readers to design their tests based on logical criteria, and Section 1.2.2.6 provides a practical example of how to use it.

1.1.2.7 Criterion for the Selection of Feedstock for AD

The theoretical concepts summarized in Table 1.1 allow us to define a selection guide based on the quality of the most common agricultural biomasses, ordered from most to least digestible, as shown in Table 1.5.

1.2 Technical Nomenclature

Each branch of science or technology has its own specialized vocabulary. Sometimes, the use of technical terms in a book or the assumption

TABLE 1.5

Anaerobic Digestibility of the Most Common Biomasses

Corn, sugar beet, bread, molasses, sugars	Easily degradable, but unsustainable
Horse dung	Digestible
Cow dung	Digestible
Goat/sheep/rabbit dung	Digestible
Organic municipal solid waste (shredded)	Digestible
Agro-food waste (vegetables, fruits, extraction cake)	Digestible
Water plants (water hyacinth, duckweed)	Digestible, but not widely available
Fresh grass	Digestible
Hen dung (layers)	Low C/N ratio
Swine manure	Low C/N ratio
Slaughterhouse waste	Low C/N ratio
Corn cobs and stalks, wheat and oat straw	Digestible, high C/N ratio
Rice straw, canes, vegetal fiber, chicken dung (broilers)	Difficult digestion
Fats and soaps, hair, feathers	Slow digestion
Wood, sawdust, hooves, horns, etc.	Indigestible

that certain technical conventions are universally understood can lead to misunderstandings and disastrous results when attempting to apply the theory to practical cases.

Many words will be recurrent within this book, and so it is worth dedicating a few lines to defining them to avoid misunderstandings.

1.2.1 Definitions

The following lines provide comprehensive definitions of the various concepts and parameters essential for the effective management of the biogas plant. The presentation order follows the logic of the forthcoming chapters. Chapter 7 contains a summary of the definitions, presented in alphabetical order, as a glossary for easier consultation.

Substrate

Biomass, animal or vegetal, that will be employed as feedstock for AD.

Inoculum

Biomass containing living microorganisms, necessary to start and maintain the AD process.

Hydraulic Retention Time (HRT) or Solids Retention Time (SRT)

It is the average amount of time that a particle or small portion of matter remains inside the digester. By definition, HRT = Volume of digester (m^3)/feedstock's flow (m^3/day).

Cellular Retention Time (CRT)

It is the average amount of time that bacteria, or a colony of microorganisms, remain inside the digester. The CRT does not necessarily coincide with the SRT and depends on the type or family of digester.

WW = wet weight, aka WM = wet mass

It is the weight (mass) of a sample "as it is" when starting a test, i.e., including its moisture.

Total solids (TS), aka dry mass (DM)

By definition, TS = WW – Moisture

VS = volatile solids

Fraction of the TS that is assumed to be digestible. By definition, VS = TS – Ash content.

Ash

The mineral fraction contained in the organic matter.

Chemical Oxygen Demand (COD)

An indirect measure of the organic carbon quantity present in the biomass. The COD is a highly standardized, relatively quick chemical test suitable for assessing dissolved volatile organic matter (e.g., alcohols, VFA).

Biological Oxygen Demand (BOD)

A measure of the organic carbon quantity present in the biomass that can be aerobically degraded. This parameter is seldom employed in the biogas industry because the aerobically degradable organic matter is not necessarily anaerobically degradable, and vice versa.

Organic Load (OL) or Total Organic Load

The total daily mass of VS or COD introduced in the digester (kg or ton of VS or COD/day)

Organic Loading Rate (OLR)

The OL divided by the digester's volume (kg or ton of VS or COD/m^3 digester · day)

Biochemical Methane Potential (BMP)

It is the net amount of methane produced by the fermentation of a mass unit of substrate in a given time. In general, the acronym BMP is followed by a suffix, indicating the duration of the test with which the BMP value was calculated. For instance, BMP_{20} refers to the total amount of methane produced during a 20-day test, per unit of mass. If BMP is not followed by a suffix, it is conventionally assumed that the test lasted 30 days. The correct unit for expressing the BMP is Ncm^3/g of VS or dm^3/kg of VS, or Nm^3/ton VS. N.B.: COD can also be employed instead of VS, especially when dealing with liquid substrates. Much of the literature expresses the BMP in cm^3/g VS or NL/kg VS. Please note that, from a formal point of view, this is not correct because liter and its subunits are not S.I. units (Système International d'Unités). Scientific papers and technical documents should report values in SI units. In Europe, it is customary to express the BMP in Nm^3/ton TS or Ncm^3/g TS or even Nm^3/ton WW. This is not correct because the moisture and ash content can vary significantly from one biomass source to another.

NL or Nm³

Volume occupied by a gas, in liters or m^3, under "normal conditions". In this context, "normal" means gas at 0°C and atmospheric pressure equal to 101.3 kPa.

SL or Sm³

Standard cubic liters or meters. This convention of defining gas volumes is widely diffused in the natural gas industry but seldom employed in the biogas industry. In this context, "standard" means m^3 of gas at 15°C and 101.3 kPa. The conversion factor between the two reference conditions is: 1 Sm^3 = 1.055 Nm^3 (conversely, 1 Nm^3 = 0.948 Sm^3).

Alkalinity (aka buffer capacity)

It is a measure of the resistance to the variations of pH of a given solution. The alkalinity is proportional to the concentration of carbonates and bicarbonates. There are several ways to express the buffer capacity; it is usually measured as the equivalent amount of CO_3Ca in mg/L of solution.

pH = hydrogen potential

It is a measure of the activity of the OH radical in aqueous medium, on a scale from 1 to 14. If pH = 7, the solution is neutral (distilled water), if pH < 7 the solution is acid (e.g., vinegar, beer, Coca Cola, lemon juice, hydrogen peroxide), and if pH > 7, the solution is said to be basic or alkaline (e.g., solution of caustic soda in water, tap water, sea water, ammonia, bleach, dishwasher soap).

Volatile Fatty Acids (VFA)/Total Alkalinity ratio (TA), aka FOS/TAC in the German and most of the European literature

Ratio between the total concentration of VFA (FOS in German) and the total alkalinity (total inorganic carbonates, TAC in German).

Lower Heating Value (LHV), aka Lower Calorific Value (LCV)

The LHV of pure methane is 9.94 kWh/Nm^3 (35.784 kJ/Nm^3), whereas for biogas, it is conventionally assumed to be 60% of this value.

Sludge, slurry, and digestate

From a scientific point of view, slurry and sludge are the same thing—the organic matter under fermentation contained in the reactor during the AD process.

Digestate is the slurry resulting from the AD process. In a single-reactor biogas plant, the sludge is simultaneously the digestate, whereas in a multiple-reactor biogas plant, the digestate is the sludge exiting from the last digester or storage tank.

Volatile Fatty Acids

Any organic acids (carboxylic acids) having fewer than six carbon atoms, i.e., formic acid (CH_2O_2), acetic acid ($C_2H_4O_2$), propionic acid ($C_3H_6O_2$), butyric acid ($C_4H_8O_2$), and valeric acid ($C_5H_{10}O_2$). The most relevant ones for the AD process are acetic acid (the preferred feedstock for acetoclastic archaea), propionic acid, and butyric acid. The presence of formic and valeric acids beyond specific concentrations is an indicator of imbalances of the microbial ecosystem or eventual problems in the AD process, because these acids should not form under normal conditions of the organic matter degradation.

Oxidation Reduction Potential (ORP), aka Redox Potential

The ORP is defined as "the tendency of a chemical species (molecule or categories of molecules or radicals) to gain electrons and hence to suffer a reduction". The ORP's measure unit is millivolt (mV). It is a chemical magnitude that is very simple and economical to measure using electrodes, which are very similar to those employed for pH measurements. The ORP is primarily used as a quality parameter for water quality in the environmental protection and wastewater treatment industries. Values between 0 and +2000 mV indicate aerobic activity. For example, values between 0 and 150 indicate a high bacterial activity consuming O_2, and hence contaminated water; values between 400 and 450 mV indicate good quality of water for a swimming pool; values from +600 mV indicate disinfection and potable water; sterilization of water with ozone brings the ORP values to over +800 mV. Values between 0 and −2000 mV indicate anaerobic processes. In the methanogenic AD process, the production of carbon dioxide is the result of an *oxidation* process of the organic carbon. In contrast, the synthesis of CH_4 is a *reduction* process of the organic carbon. Hence, we can deduce that, during anaerobic fermentation, the reactions are incredibly complex to monitor, since they encompass both oxidation and reduction of the organic carbon present in the digester. The literature indicates typical ORP values between −220 and −290 mV as those optimum to maintain a stable AD process (Blanc and Molof, 1973), but ORP is not a reliable parameter (discussed in detail in Section 1.4.6).

1.2.2 Practical Applications and Numerical Examples

It is helpful to dedicate some time to becoming familiar with the definitions provided in Section 1.2.1, as they will be frequently referenced throughout this book. The reader is encouraged to review the exercises proposed in the following sections, having a paper and pen in hand.

1.2.2.1 Approximate Ratio between COD and VS

In many cases, it may be beneficial to compare the results of our tests with those reported in the scientific literature. Unfortunately, there are no strict norms about the measurement units in which the values must be expressed. One recurrent case is the use of COD or VS as different ways to estimate the potentially degradable organic matter or the amount of living microbes contained in a sample. From a formal point of view, there is no difference between adopting COD or VS as measure units of the organic matter concentration, since both are indirect indicators of the quantity of carbon contained in a given biomass. From a formal point of view, the only univocal parameter to assess the organic matter content of a biomass sample is its TOC (total organic carbon), but it is difficult and expensive to measure and, being the result of a difference between two measurements, the uncertainty is high (see Section 2.1.2), so that's why TOC is never employed in the biogas industry. Regardless of whether you choose VS or COD, the important thing is to maintain the coherence of the measurement units when performing the calculations. Generally, COD is the preferred unit to measure the organic matter contained in liquid substrates (sludge, wastewater, whey, etc.) for two reasons: because it is easier to perform a COD test on a liquid, and because volatile organic matter (alcohols, aldehydes, and VFA) will be wrongly counted as water. VS, on the other hand, are easier to measure when analyzing solid biomasses. The particular case of silage, which contains both solid organic matter and VFA, will be dealt with in detail in Section 3.2.1.4.

If the sample contains no VFA, it is possible to establish a reasonably approximated theoretical correlation between COD and VS, assuming that 50% of the VS mass is made of C (N.B.: this is relatively true for most biomasses, especially vegetal ones, but not applicable to *all* substrates!).

From stoichiometry, it is known that 32 g of O_2 is necessary to oxidize 12 g of C completely.

Hence, the COD of a given quantity of VS is equal to:

$$\mathrm{COD} = 0.5 \cdot \mathrm{SV} \cdot \frac{32}{12} = 1.333 \cdot \mathrm{SV}$$

Please note that VS can be expressed as % of the WW or as % of the DW

The convention adopted in this book expresses the VS as % of the WW.
COD, in turn, is usually expressed in mg of O_2/L of slurry or liquid substrate (in rare cases, also as g of O_2/g of dry weight, DW).

EXAMPLE NO. 1

Calculate the COD of 60 g of Sludge having 4% of VS

$$VS_{sludge} = 0.04 \cdot 60\ g = 2.4$$

$$COD_{sludge} = 1.333 \cdot 2.4\ g = 3.2\ g\left(\text{N.B., these are absolute g of } O_2\right)$$

$$COD_{sludge} = 3200\ \text{mg/60 mL} = 53.333\ \text{mg of COD/L}$$

Please note another convention, which is not a written norm but is routinely accepted worldwide: in general, it is assumed that sludge and liquid substrates have the same density as water; hence, the 60 g of our sample is equivalent to 60 mL. In reality, sludge is denser than water—from 1020 to 1060 g/L—but this difference can be neglected in calculations for industrial purposes.

The correlation between COD and VS found earlier is not valid for all organic matters, as the following practical example shows.

EXAMPLE NO. 2

Calculate the COD of a 10 g sample of pure glucose.

The total oxidation of glucose is described by the following stoichiometric formula:

$$\begin{array}{cccc} C_6H_{12}O_6 + & 6O_2 \rightarrow & 6CO_2 + & 6H_2O \\ 180\ g & 192\ g & 264g & 108g \end{array}$$

Hence, 180 g of VS (since glucose has no ash, i.e., it is fully volatile) is equivalent to 192 g of COD. Hence the equivalence ratio is 1 g COD = 1.0666 g of glucose.

From the same reasoning as the practical examples 1 and 2, it is possible to derive the conversion Table 1.6, which will prove immensely helpful when performing tests with pure substances employed as reference substrates.

TABLE 1.6

Equivalence Ratios between VS and COD of Some Pure Substances Employed for Performing Laboratory Tests in Biogas Plants

Substance	Chemical Formula	g COD/g VS (or anhydrous liquid product)
Vegetal and bacterial biomasses	≈50% C	1.333
Glucose	$C_6H_{12}O_6$	1.067
Sucrose (saccharose)	$C_{12}H_{24}O_{12}$	1.067
Fructose	$C_6H_{12}O_6$	1.067
Starch	$(C_6H_{12}O_6)_n$	1.067
Cellulose	$(C_6H_{10}O_5)_n$	1.185
Proteins	Empirical formula $C_{400}H_{620}N_{100}O_{120}P_1S_1$	2.170
Acetic acid	CH_3COOH	1.067
White vinegar (6% acetic acid)	6% CH_3COOH + 94% H_2O	0.064
Sodium acetate	CH_3COONa	0.780
Propionic acid	$C_3H_6O_2$	2
Sodium propionate	$C_3H_5O_2Na$	1
Butyric acid	$C_4H_8O_2$	2
Sodium butyrate	$C_4H_7O_2Na$	1.6
Isopropanol	C_3H_8O	2.4
Vegetable oils (assumed as oleic acid)	$C_{18}H_{34}O_2$	2.89

1.2.2.2 Methane Yield of a Substrate with a Known BMP

Generally, the scientific literature expresses the BMP of the substrates in terms of VS or COD. This is the correct mode to obtain universally comparable test results. The reason is that a given substrate may contain varying amounts of moisture and ash, depending on an infinite number of factors. The substrate's BMP will be directly proportional to the VS percentage and should not diverge much from a characteristic figure, typical for the substrate in question. Many agricultural magazines and most suppliers of AD feedstock express the BMP either as a function of the WM or as a function of the DM. Both criteria are somehow treacherous because two different lots of a given substrate, e.g., silage, may have the same percentage of dry mass, but their VS percentage (and consequently their BMP) can be very different for several reasons. The following examples, drawn from real cases, illustrate how to perform the correct calculations in each instance, thereby obtaining coherent and comparable results.

EXAMPLE NO. 3

A source in the scientific literature states the following values for corn silage:

TS = 30.7%
VS = 95.5% (of TS)
BMP = 310 NL/kg VS

An agricultural magazine quotes the following data for the same substrate:

TS = 30%
VS = 91.5% (on TS)
Biogas yield = 668 m^3/ton TS

Is it possible to compare both sets of values?

Please note that the magazine adopts an incorrect criterion, as it defines the yield in terms of biogas but does not specify the percentage of CH_4 the biogas contains. It is conventionally assumed that biogas is composed of 60% CH_4 and 40% CO_2. Under industrial plant conditions, or for some special pure substrates, the proportions of CH_4 and CO_2 are highly variable. Corn silage typically yields biogas with approximately 55% CH_4. Furthermore, the author of the article does not specify whether the gas volume is expressed in Nm^3 or Sm^3.

Suppose, as a first approach, that the author of the article has employed the most frequent conventions of the biogas industry. Replacing the TS by the usual equivalent VS content, we obtain:

$$\text{Biogas} = \frac{668\left[\dfrac{Nm^3}{t\,TS}\right]}{0.915\left[\dfrac{t\,VS}{t\,TS}\right]} = 730\left[\text{Nm}^3/\text{t VS}\right]$$

$$\text{BMP}_a = 0.6 \cdot 730\left[\text{Nm}^3/\text{t VS}\right] = 438\left[\text{Nm}^3/\text{t VS}\right]$$

The BMP value thus obtained appears too high when compared to the typical BMP found in the scientific literature that we are using as a benchmark. Let us recalculate the BMP assuming that the biogas contains only 55% CH_4 (the usual composition for substances composed mainly of carbohydrates, such as silage). Under such supposition, the BMP deduced from the magazine's article will be nearer to the value found in the scientific literature.

$$BMP_b = 0.55 \cdot 730\left[Nm^3/tSV\right] = 401\left[Nm^3/tSV\right]$$

If the author of the article normalized the volume in Sm^3, then the resulting BMP expressed in the correct units is:

$$BMP_c = \frac{0.55 \cdot 668\left[\dfrac{Sm^3}{tTS}\right] \cdot 0.947\left[Nm^3/Sm^3\right]}{0.951\ [t\,VS]} = 380\left[Nm^3/tSV\right]$$

By reviewing the literature, it will appear that, depending on the authors of different studies, the BMP of corn silage varies from 338 Nm^3/ton VS to 537 Nm^3/ton VS. So, who is right?

We leave the reader to meditate on this apparent contradiction, as we will address the argument of error propagation in determining the BMP of silage in the forthcoming chapters. The essential concepts to keep in mind from this chapter are:

1. The BMP that can be effectively obtained from a given biomass in a given biogas plant is not a universal and absolute value.
2. Not all bibliographical sources are reliable.
3. It is crucial to pay attention to the units in which the BMP is expressed, especially when the numerical values are very high, and even more especially when said values are published in "nonscientific" magazines, because quite often such articles are sponsored by companies or stakeholder groups having a particular interest in showing high productivity of their products.

1.2.2.3 Coherent Use of the TS, VS, and BMP Values

EXAMPLE NO. 4

Upon measuring the TS, VS, and BMP of a silage sample taken from our stock, we determine 31% TS, 29% VS (both expressed on WM), and 348 Nm^3/ton VS. How many tons of silage should we feed to the digester to bring the plant to work at its nominal power of 500 kW? According to the manufacturer's technical sheet, the generator's electric efficiency is 38%.

First, we need to calculate the primary energy necessary to feed the generator at full power during 24 h.

$$E = \frac{500[\text{kW}]\cdot 24[\text{h}]}{0.38} = 31{,}578.94[\text{kWh}]$$

Given the LCV of methane, its necessary daily quantity, *G*, will be:

$$G = \frac{31{,}578.94[\text{kWh}]}{9.94[\text{kWh/Nm}^3]} = 3{,}177[\text{Nm}^3]$$

The total OL to feed the digester will then be equal to:

$$\text{OL} = \frac{3{,}177[\text{Nm}^3/\text{d}]}{348[\text{Nm}^3/\text{ton VS}]} = 9.13[\text{ton VS/d}]$$

Since the silage has 29% of VS on WM, the total daily amount of silage to load on the plant's hopper, *M*, will be:

$$M = \frac{9.13[\text{ton VS/d}]}{0.29} = 31.5[\text{ton/d}] \tag{1.13}$$

N.B.: This is the typical (linear) calculation adopted by most biogas plant managers. Please note that this simple calculation assumes the total degradation of the silage (i.e., extracting its whole BMP) and a constant production of biogas directly proportional to the OL, which is not true in most cases.

1.2.2.4 Calculation of the HRT

EXAMPLE NO. 5

Consider the same data as in the former example no. 4. Assuming that the plant consists of a single digester with 4000 m^3 of useful volume and that, to feed it through a pump, the silage must be diluted with 4 m^3 of water per ton of WW. What are the HRT and the SRT?

In this case, the HRT is coincident with the SRT, since we suppose that the digester is perfectly mixed. Such an assumption is generally valid for most existing plants, but it is not an absolute truth. We also assume that all solids will be converted into biogas during the digestion process (although this is not entirely accurate, it is a reasonable assumption for highly degradable matter, such as corn silage).

The total daily flow of liquid loaded to the digester, Q, results from the following formula:

$$Q = 31.5\left[\text{ton/day}\right]\cdot 4\left[\text{m}^3/\text{ton}\right] = 126\left[\text{m}^3/\text{day}\right]$$

By definition of HRT, we have:

$$\text{HRT} = \frac{V}{Q}$$

where

V = useful volume of the digester [m^3]

Q = hydraulic load [m^3/day]

Hence HRT = 4000/126 = 31 days

N.B.: Please note that the useful volume of the digester is never equal to the geometric volume, because a certain fraction of the total volume must be left free for the eventual formation of foam on the slurry's surface; hence, it is impossible to fill a digester to its maximum (geometric) capacity.

1.2.2.5 Checking the OLR

EXAMPLE NO. 6

Calculate the OLR of the digester considered in the former examples. By definition:

$$\text{OLR} = \frac{\text{OL}}{V}$$

Hence

$$\text{OLR} = 9130\ [\text{kg VS/day}]/4000\ [\text{m}^3] = 2.3\ \text{kg VS/m}^3 \cdot \text{day}$$

In a conventional biogas plant (stirred reactor, average feedstock solids content in the range of 9%–12%), the OLR should remain in the range of 2–3 kg VS/m^3 · day (it is advisable to keep it nearer to 2 than 3 for maximum efficiency of the conversion of biomass to methane). Hence, the digester is correctly loaded.

1.2.2.6 Optimizing the C/N Ratio of a Substrate Mixture

EXAMPLE NO. 7

In a farm with 100 pigs, the daily production of manure is equal to 1000 kg/day. Oat straw is available as a cosubstrate. What is the quantity of straw that the farmer should add to the manure to keep the biogas plant working at its optimum capacity?

Data (see Table 1.4)

- C/N of swine manure = 4%/0.4% = 1000·(0.04/0.004) [kg/day]
- C/N of oat straw = 50%/0.5% = (0.5/0.005)·x [kg/day]
- Ideal C/N ratio of the mixture for optimum digestion = 150/5

The amount of straw, x, necessary to obtain a mixture with ideal C/N ratio is hence:

$$\frac{150}{5} = \frac{1000 \cdot 0.04 + x \cdot 0.50}{1000 \cdot 0.004 + x \cdot 0.005} = \frac{40 + x \cdot 0.50}{4 + x \cdot 0.005}$$

$$150 \cdot (4 + x \cdot 0.005) = 5 \cdot (40 + x \cdot 0.50)$$

$$600 + x \cdot 0.75 = 200 + x \cdot 2.50$$

$$600 - 200 = x \cdot (2.50 - 0.75)$$

$$x = 228.6 [\text{kg/day}]$$

1.3 Managing the Plant "by Tables": Limitations and Risks

Most industrial applications employ two different technologies of process control, called "open loop" and "closed loop" controls. We illustrate the meaning and advantages of each one, as well as their limitations, with a practical example. The specific literature on the theory of control systems is provided for readers who wish to gain a deeper understanding of the topic.

Suppose we desire to heat a room with an electric stove. There are two ways to control the flux of energy to the stove:

1. By connecting the stove to the electric supply through a timer-controlled switch. In such a case, the timer activates the heater and maintains it running for a predefined period, during which the room's temperature will reach a specific value. For sure, the final value will be higher than the initial one, but we cannot know it in advance, as it will depend on the boundary conditions (initial temperature of the room, thermal dispersion of the room's envelope, potential voltage variations while the timer is *on*, etc.).
2. By installing a temperature sensor and a thermostat in the room, which will control a relay that turns the stove on and off. In such a system, the electric stove will heat the room until its temperature reaches the desired value set in the thermostat. The time to reach the preset value will depend on many factors, as in case (1), but the system (2) has the advantage of reaching exactly the temperature desired by the user, regardless of any other circumstance.

A system like the one described in Example 1 is known as an open-loop control system. We could well call it "dumb control system" because, by starting the timer, the electric heater will turn on regardless of the effective temperature of the room and will heat it during a time, predefined by the manufacturer with some general criterion, unknown to us, which does not necessarily apply to our specific case. By adopting such a control system, the user will most of the time either waste energy by overheating the room (in cases where it is initially warm) or fail to reach a comfortable condition during icy weather, when the preset time may not be enough to bring the temperature to an acceptable value.

A heating control like the one described in Example 2 is undoubtedly better, both from the energy consumption and comfort points of view, because it turns the heater on and keeps it on only until the desired temperature is reached.

Both control techniques are usually depicted in the technical literature with the diagrams shown in Figure 1.5.

Despite the industrial biogas plant costing millions of euros, closed-loop control systems are employed only for the temperature of the digesters and the motor unit. In contrast, the management of the biological process is mainly based on open-loop logic.

In practice, most biogas plant managers rely on tables of BMP, pH, VFA/TA (FOS/TAC), ORP, electric conductivity, and similar parameters to decide how to manage their plants. The experience shows that the said tables are seldom reliable, as they come from different bibliographical sources, published by researchers who have performed measurements under conditions that rarely apply to our plant, our feedstock, or both.

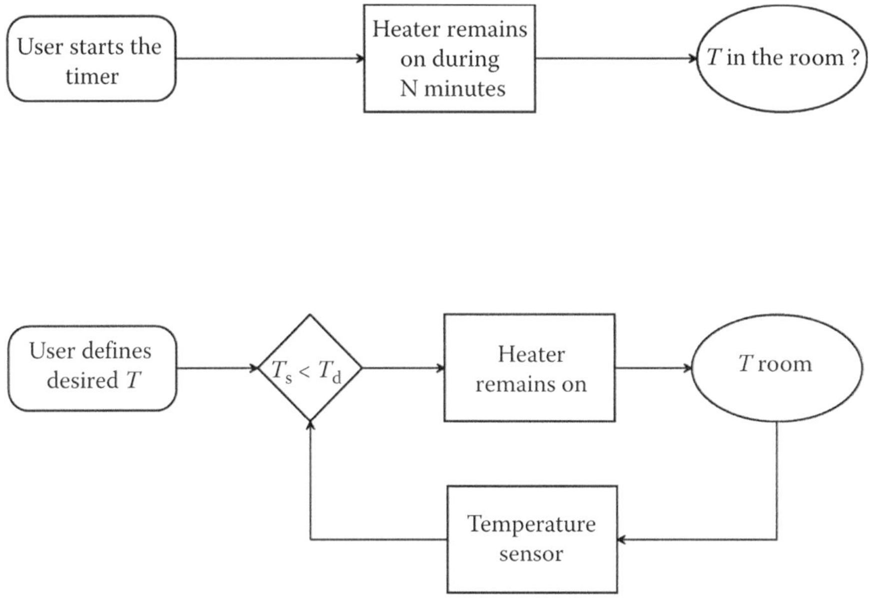

FIGURE 1.5
Block diagrams of an open-loop control (above) and a closed-loop control (below).

The appealing concept underlying the use of tabulated values to manage a biogas plant is the linearity of the approach: the plant manager hopes to achieve the desired methane production simply by performing some straightforward calculations based on values that have been determined in advance. Such an approach quite often results in producing less energy than expected, while in other (rare) cases, it may lead to methane overproduction, which means wasting both energy and feedstock. The most widely adopted approach in the biogas industry involves maintaining production at its nominal maximum, albeit at the expense of sacrificing efficiency in converting biomass into methane. In any case, the economic result will be less than optimal. When underfeeding the plant, the biomass is digested with high efficiency, but the energy production will be less than expected. In the opposite case, the nominal power of the plant is reached, and the energy production is kept constant, at the expense of consuming more feedstock than expected. From an environmental point of view, overfeeding results in emitting more greenhouse gases into the atmosphere (higher residual methanogenic capacity of the digestate, due to the incomplete digestion of the feedstock) and requires managing a bigger quantity of digestate. For these reasons, we can conclude that managing the biogas plant based solely on tabulated values is attractive due to its apparent simplicity, but it is neither effective nor economically efficient.

Many biogas plant manufacturers propose control solutions to manage "automatically" the plant, apparently based on the closed loop concept, but it is necessary to point out that the theorems on which the closed loop control theory is founded apply only to *linear systems*, i.e., each action will produce a directly proportional reaction. Closed-loop systems can, of course, control nonlinear processes, provided that the control logic is also nonlinear, thereby making it more complex. Unfortunately, the AD is a nonlinear process. We can even say that it is "complicatedly nonlinear". The mathematical function that better describes the degradation of organic matter and its corresponding transformation into methane is indeed a complex exponential, as shown in the following examples.

1 Example of a linear equation (for instance, the amount of natural gas delivered by a tube connected to a constant pressure grid during a certain time):

$$H = a \cdot t$$

where

H = total amount of gas cumulated in time t

a = a proportionality coefficient depending on the grid's pressure and the tube's length and section

2 Mathematical model of the AD (aka the *Gompertz equation*). Its graphical representation is a curve called *sigmoid* because of its shape resembling an *S*:

$$H = a \cdot e^{\left[-b \cdot e^{-ct}\right]}$$

where

H = methane production cumulated in time t

a = upper asymptote of the curve, physically corresponding to the substrate's BMP

e = base of the natural logarithms = 2.71828…

b = a coefficient that is proportional to the acclimation time required by the bacteria (called *lag phase* in technical literature), physically meaning the necessary time to reach the maximum daily flow

c = a coefficient that is proportional to the maximum daily flow

t = elapsed time (the variable of the Gompertz equation, usually expressed in days)

Another form of the Gompertz equation usually found in the literature is the following:

$$H = a \cdot e^{-\left[\frac{c \cdot e}{a}(\lambda - t) + 1\right]}$$

where

H = methane production cumulated in time t

a = upper asymptote of the curve, physically corresponding to the substrate's BMP

e = base of the natural logarithms = 2.71828…

λ = lag phase (days)

c = maximum daily flow (cm^3/day)

t = elapsed time (the variable of the Gompertz equation, usually expressed in days)

For the reader who is not familiar with exponential functions, it is interesting to graphically compare the proportional (linear) process (i.e., the assumed behavior when applying data from tables) and the sigmoid curve, which is the real-life case (Figure 1.6).

When managing a biogas plant, it turns out to be useful to know the total methane flow that a given quantity of biomass can yield, as a function of time. By definition, the flow is the quantity of gas produced in a unit of time, a quotient that can be calculated at each point of the cumulative volume curve by dividing the latter into very small intervals. In Mathematics, the operation in question is called *differentiation*, and the curve thus obtained is referred to as the *derivative*. The derivative of a sigmoid as a function of time

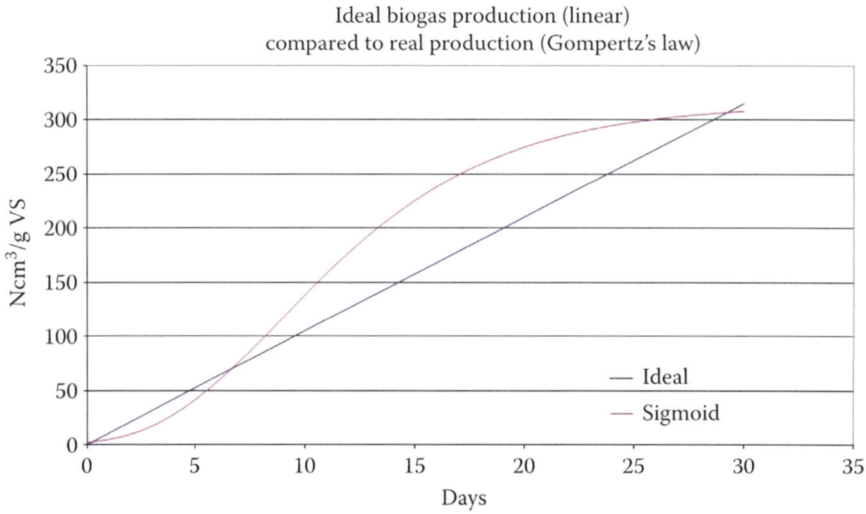

FIGURE 1.6
Cumulative methane production from the AD of organic matter: real biogas production compared to the assumption of the linear output.

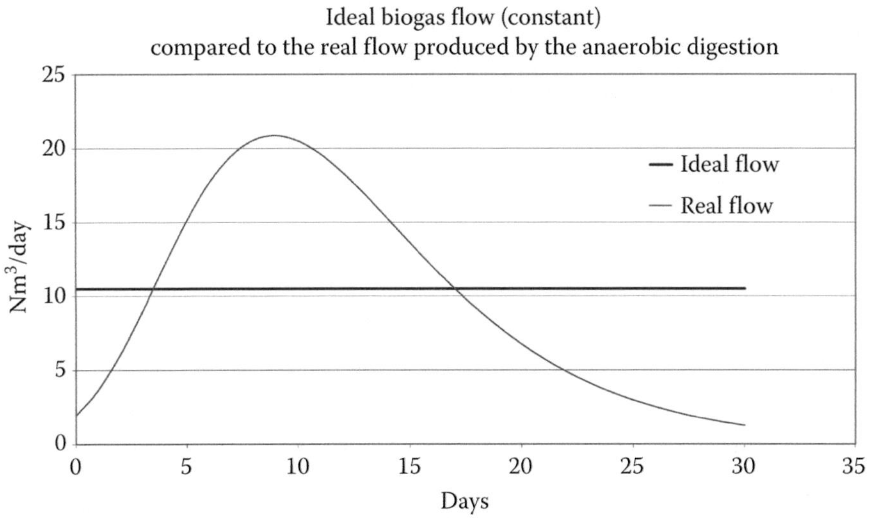

FIGURE 1.7
Daily biogas (or methane) flow as obtained from a batch test.

(hence the curve representing the instant flow at any time) is a bell-shaped curve, represented by the following mathematical expression:

$$H' = a \cdot b \cdot c \cdot e^{-b \cdot e^{-c \cdot t}} \cdot e^{-c \cdot t}$$

Figure 1.7 clearly illustrates the substantial difference between the "ideal" behavior that any biogas plant manager desires (linear degradation of the biomass to obtain a constant biogas flow) and the actual behavior of the AD of any substrate.

In practice, we need to apply the results of a batch test on a biomass sample (the sigmoid) to a plant that must produce continuously. How can we utilize the results of the laboratory test to determine the diet of our biogas plant? To answer this crucial question, we must consider that the ultimate scope of all the tests described in advance is to determine how much of a given feedstock must be fed to the digester to achieve the desired (average) methane flow. Flow is, by definition, the quotient of the gas volume (or the mass) by the unit of time. If we divide the sigmoid curve into intervals of 1 day and tabulate the quantity of methane produced in each interval, we will obtain the curve of daily gas flow, as shown in Figure 1.7.

Fortunately, a continuous digester—i.e., a digester receiving a certain amount of feedstock each day—can be envisioned as a group of batch digesters being fed sequentially. Hence, each one will generate the same daily flow curve, but each curve will be delayed by 1 day compared to its predecessor, as shown in Figure 1.8.

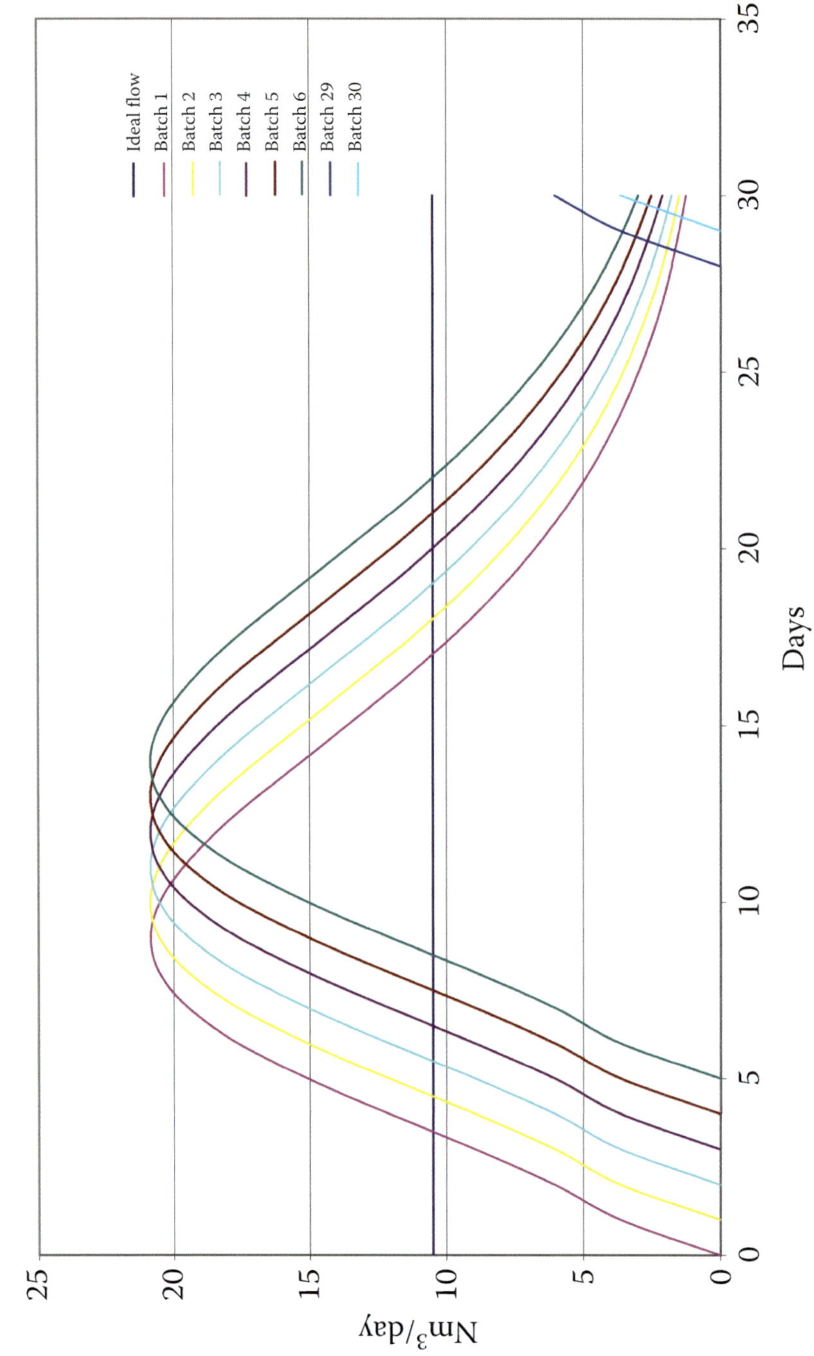

FIGURE 1.8
Simplified mathematical representation of a continuous digester as a group of sequentially fed batch digesters.

Hence, under the given supposition, the total biogas flow produced by a continuous digester fed a certain daily amount of biomass, *M*, is equivalent to the sum of the instant flows of *n* batch digesters fed the same amount *M*, once every *n* days. The result of this operation, assuming that the plant is already working in steady state, is shown in Figure 1.9.

The former reasoning and the result shown in Figure 1.9 both explain three essential facts:

1. Now we can understand why most biogas plant constructors adopt the Continuously Stirred Reactor Tank (CSRT) technology. As can be seen from Figure 1.9, the said reactor type "linearizes" the behavior of a process that is not linear itself.
2. It should be clear now why, in continuous biogas plants, it is necessary to feed a daily quantity of substrate slightly higher than the one calculated as the simple quotient between the daily necessary methane quantity and the BMP. The reason is that the anaerobic degradation of any substrate has an asymptotic limit. Hence, in practice, it is impossible to extract 100% of the feedstock's BMP. The digester's efficiency (the effective amount of gas produced in a certain time compared to the BMP measured in batch tests in the same time) can reach high values. Still, it cannot reach 100% unless

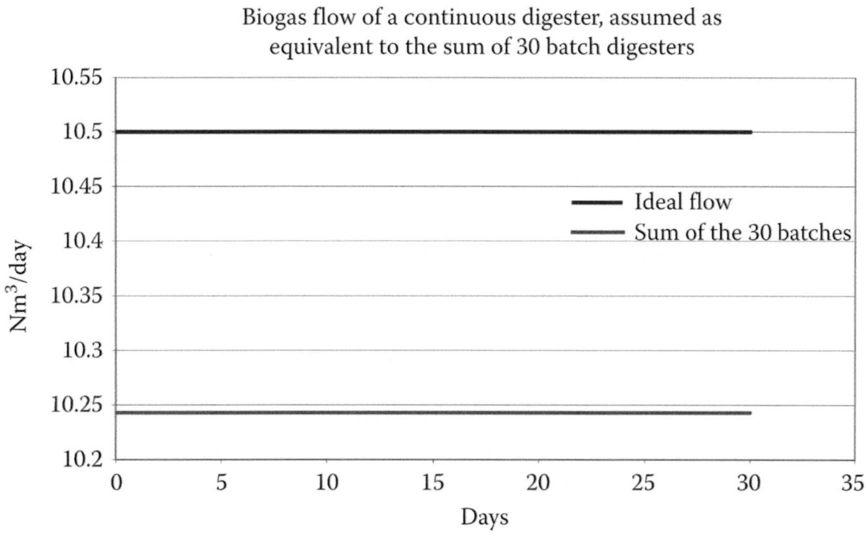

FIGURE 1.9
Real flow from a continuous digester with 30 days retention time, assumed as a group of 30 batch digesters, each one being loaded every 30 days with the same amount of feedstock that constitutes the daily ration of the continuous digester (violet line), compared to the flow calculated with the simple linear assumption, BMP/30 days (blue line).

the HRT (or SRT) is long enough (theoretically infinite in the mathematical model of the Gompertz equation). A second reason, often overlooked by biogas plant designers and forgotten by academicians, is that a digester is essentially a microbial farm. Like any other living beings, microorganisms grow and reproduce. Consequently, part of the feedstock contributes to increasing the mass of living bacteria in the sludge, rather than becoming CH_4 and CO_2.

3 The constructors of biogas plants program their control systems based on a "closed loop" logic, which relies on BMP values "from tables". The BMP is represented by the parameter *a* in the Gompertz formula, but the average biogas flow also depends on the coefficients *b* and *c*, which, in general, are not tabulated. Even in the case a biogas plant constructor should consider the values *a, b*, and *c* as variables for programming his control system, the said values depend not only on the digester's feedstock but also on the state of the bacterial system (in the following chapters we will define in detail what *state* means in our context). Hence, the control systems sold by most biogas plant constructors as "technologically advanced" (because based on a "closed loop" logic) are only partly effective. In practice, they contain a certain extent of the "open loop" logic, since their operation has been programmed in the factory on the basis of coefficients "from tables", and not necessarily coincident with those really describing the features of the feedstock and the activity of the bacterial consortium in a given moment.

1.3.1 Conclusions on the Use of Tables

The limitations of the biogas plant management arising from the use of tables are two:

1 The tables only show *one parameter* of the whole AD curve, the BMP, which in turn is highly variable as a function of several factors, like the chemical composition and grain size of the feedstock.

2 The other two parameters describing the AD dynamics are impossible to tabulate because they depend on the biodiversity and activity of the inoculum, the process temperature, the pH and alkalinity of the mixture under digestion, the reactor's geometry, the stirring intensity, the presence or absence of biocatalysts or inhibitors, and so on.

The problem is further complicated when the digester's diet is composed of mixtures of organic matter, since the BMP of a mixture is not necessarily the arithmetic sum of the individual BMPs. The explanation for this apparent paradox is that the anaerobic bacterial ecosystem thrives at its best when the C/N/P ratio is nearly 150/5/1. With this in mind, it appears evident that mixing two substrates, for instance, both rich in N, will bring down the C/N ratio, creating in some cases a partial inhibition of the whole process, because of the defect of carbon and excess of ammonia. Conversely, a mixture of two substrates in such proportions that the resulting C/N ratio is nearly 30 could produce a synergic effect, producing more methane than the arithmetic sum of the single BMPs measured separately.

A table of the BMPs of several common substrates is included just as an example.

Surely, the reader is already disagreeing with some of the said values, since they may differ too much from the values that can be deduced from the plant's operation. Feedstock suppliers and plant constructors may declare very different values, too. Which value is then correct? The same perception that something was wrong when comparing BMP values of the same feedstock has driven the scientific community to question whether the test protocols for the BMP assay were wrong, given the high variability of the results found by different research groups.

Many experiments have been conducted to find out the reason for the large discrepancies between the BMP values measured by different laboratories. For instance, in 2011, Raposo et al. (2011b) conducted a study on 18 laboratories, who had to determine the BMP of a 100% degradable and pure substrate whose theoretical BMP is well defined: starch. Figure 1.10 shows the result.

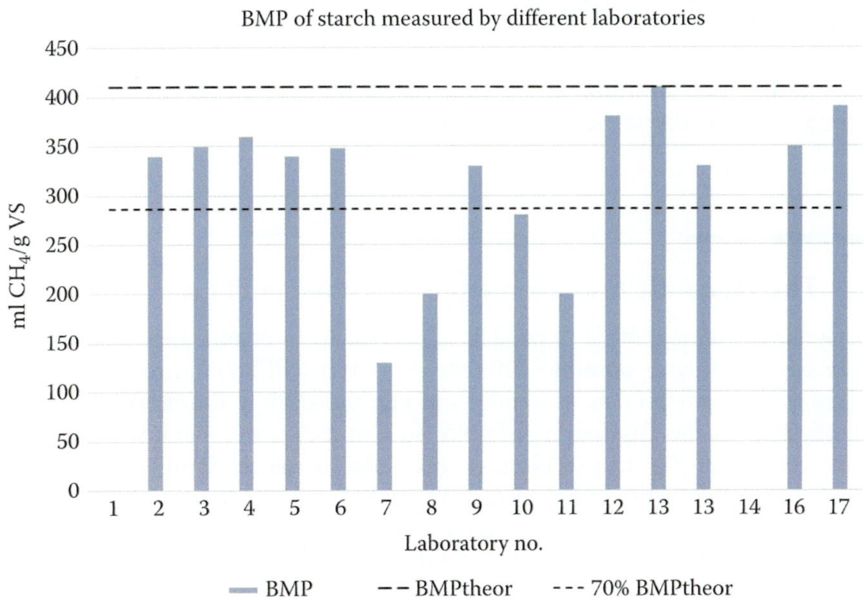

FIGURE 1.10
The BMP of starch, as measured by 18 different laboratories. (Raposo, F., et al., *Journal of Chemical Technology and Biotechnology*, 86(8), 1088–1098, 2011b; graphics by the author.)

The reader can now easily realize that, if the results are so extremely variable when measuring the BMP of a perfectly defined pure substrate, and if we discard the possibility of human errors (because the tests were performed by expert laboratory operators under controlled conditions), the causes of the variability of the results must be quite more elusive.

In the first approach, we can preliminarily conclude that the BMP does not depend only on the chemical composition of the substrate, but also on the bacteria that compose the inoculum, on the preparation of the sample, on the test conditions, etc. Even the geometry of the reactors and their stirring have a certain influence, but of second order.

After 15 years measuring the BMP of hundreds of different substrates and the methanogenic activity of inoculums samples from hundreds of biogas plants, the author can demonstrate the universal validity of Rosato's Laws, which are a consequence of Murphy's Law:

1 Under perfectly controlled laboratory conditions, a biological system will always behave as it wishes.
2 The biological system's behavior is usually the contrary of the one expected.

Jokes apart, it is perfectly valid and normal that in a given biogas plant, fed with the same biomass, with the same process parameters, and following the same routine, it is impossible to obtain a perfectly constant methane flow by just controlling the feedstock flow to the digester (unless the digester is somehow overfed, as almost everybody does). Some "closed loop control" (not necessarily an automatic device) must be applied. This objective fact brings us to the discussion in the next section.

1.4 The Dynamic Management of the Biogas Plant

A dynamic management approach is necessary to ensure the biogas plant runs as smoothly as possible. This means that it will be required to monitor as many parameters as possible, among those that influence the three coefficients in the Gompertz sigmoid (a, b, and c), then check the flow and composition of the output biogas, and hence act accordingly on the feedstock input. Unfortunately, no sensor exists (at least, not as a commercial product as of July 2025) that can instantaneously measure what is happening inside the digester or predict the methane yield of a given substrate at a given moment. It is necessary to perform the BMP and biological activity tests. The reader may have observed that even the same lot of feedstock from the same silo, in the same plant, and under the same fermentation conditions will yield different quantities of methane over time. The explanation of this

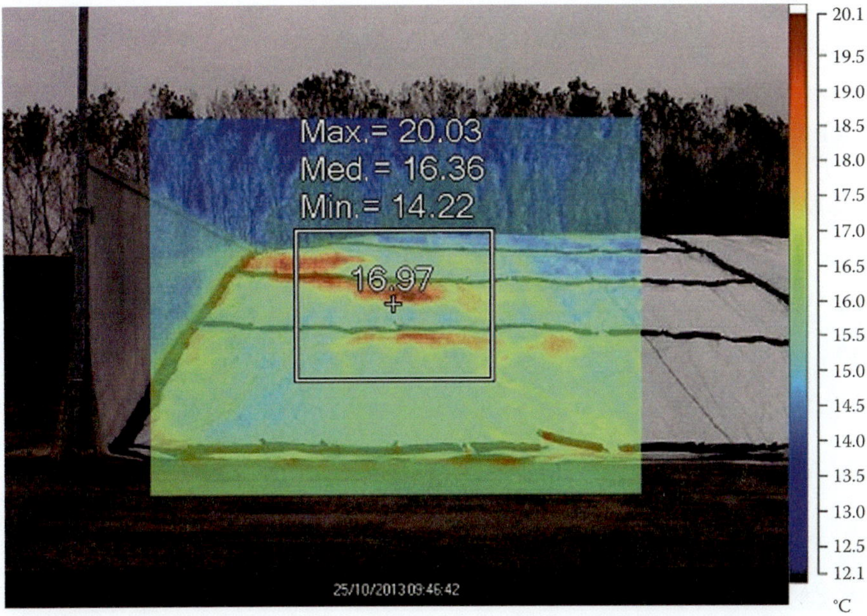

FIGURE 1.11
IR photographs of a trench filled with corn silage. (© Arch. Giovanna Barbaro, certified second-level IR thermography operator.)

phenomenon is intuitive, as the ensilage process is, in itself, a form of pre-fermentation of the vegetal biomass; however, it continues over time, albeit at a slow pace. Hence, if we perform a BMP test in the very month the biomass is ensiled, finding some given values for a, b, and c, it is certain that, after some months, these values will be different. Figures 1.11 and 1.12 show the thermal photos (taken with an infrared camera) of two trenches, both containing corn silage. The red spots indicate zones where the temperature is higher, a clear signal that aerobic bacteria are oxidizing the silage into carbon dioxide (an exothermic reaction, hence the heat detected by the IR camera). As time passes and carbon is lost to the atmosphere, less carbon will remain in the silage, so a given quantity of feedstock will produce less methane. The causes of such uncontrolled oxidation are surely known to the reader—mice and crows pierce the membrane that covers the silage, allowing air to enter; the difficulty in closing the trench's front in an airtight way each day; more or less compaction and air trapped during the ensiling operations…

Several parameters have been proposed to monitor the AD process to "close the loop" of the control system with useful information, allowing us to dynamically vary the input of biomass to the digester, obtaining at any moment the maximum methane production with the minimum biomass consumption.

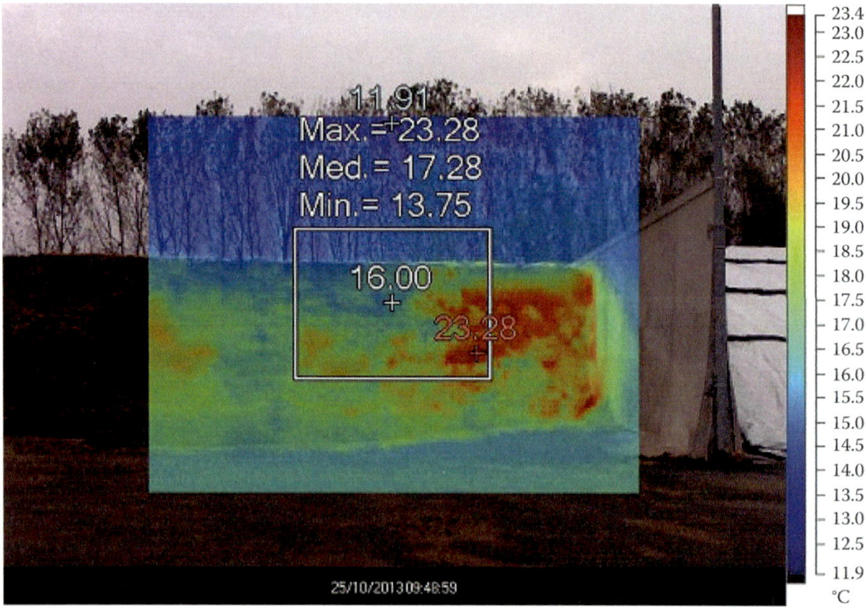

FIGURE 1.12
IR photographs of a trench filled with corn silage. (© Arch. Giovanna Barbaro, certified second-level IR thermography operator.)

From a didactic point of view, it may be useful to imagine the biogas plant as a car, which we need to drive for 1,000 km at a constant speed of 150 km/h. In the system called "car", the instruments on the dashboard and the view of the road through the front window provide the driver with the necessary information to "close the control loop", i.e., to press the gas pedal more or less. Let us analyze some of the most common parameters that comprise our imaginary "dashboard of the biogas plant", to understand their validity and why, even when combined, it is impossible to maintain the AD process at a constant rate.

1.4.1 The pH

It is well proven that the archaea thrive at their best when their environment has a pH in the range of 6.8–7.2, as can be seen from the research by Andrews and Greaf (1971), as shown in Figure 1.13, taken from Gerber and Span (2008).

The fermentative bacteria, in turn, prefer environments with pH ranging between 5 and 6.5, although in practice they can tolerate pH between 4 and 8.4. According to these facts, it appears evident that the optimization of one of the steps of the AD process will automatically hinder the correct development of the other, resulting in a lower specific production of methane for a

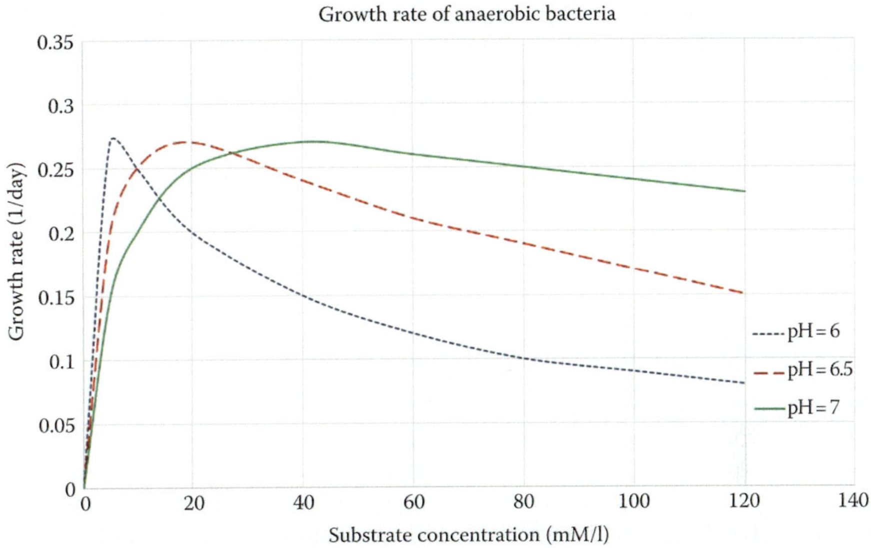

FIGURE 1.13
The reproduction speed of the *Archea* as a function of pH and substrate concentration. (Data from Andrews and Greaf, 1971; Gerber, M., Span, R., *International Gas Union Research Conference*, Paris, France, 2008; graphics by the author.)

TABLE 1.7
Indicative BMPs of Several Agricultural Substrates

Substrate	Average BMP (Nm³ CH_4/t VS)
Swine manure	450
Cattle manure	250
Horse manure	460
Sheep manure	200
Manure from stable (with straw)	225
Chicken dung	450
Corn straw	410
Corn silage	550
Rice straw	220
Grass	410
Vegetal waste	350
Sewage water	420

given digester and feedstock. According to Mata-Alvarez (2002), the inhibition caused by NH_3 is predominant at pH > 7, while the inhibition caused by SH_2 predominates at pH < 7. The production of both inhibiting compounds together with CH_4 is unavoidable with some substrates (please refer to Table 1.7). Hence, in extreme conditions, it is possible that both compounds reach toxic levels, causing the biologic collapse, even if the pH remains close to 7.

In the concrete case of digesters fed with manure, its high buffer capacity (aka alkalinity) prevents the pH from changing appreciably. When the pH eventually reaches an alarm threshold, it is because the concentrations of SH_2 or NH_3 have become so high that it is usually too late to save the digester from the biological collapse. We can conclude that pH is not a 100% reliable control parameter. In our analogy with the car's instrument panel, we can compare monitoring the pH to the oil pressure warning: it will light up when the motor is nearly, if not already, seized. Measuring the pH has the advantage of being cheap and easy.

1.4.2 The Relationship between pH and Alkalinity

Based on the previous considerations, if pH and alkalinity are measured (i.e., we know the system's resistance to pH variations), we can assess the stability margin of the biological process with higher accuracy. We have already explained that Archaea thrive when the pH remains between 6.8 and 7.2. From similar tests, it is known that the ideal alkalinity should be in the range of 8000–9000 mg equivalent of calcium carbonate per liter (mg eq. $CaCO_3$/L), as higher values may lead to inhibition and dissolution of the active granules, while lower alkalinity values will allow wider pH excursions.

In the AD process, the alkalinity is given by the presence of carbonates and bicarbonates (dissolved carbon dioxide), ammonia, phosphates, silicates, VFA, and the total alkalinity is usually high (in the order of several g equivalent of carbonate/L of slurry).

The constructors of biogas plants that employ the alkalinity adjustment as a control method (only a few, since this is an obsolete technique) install automatic dosing systems of sodium carbonate or bicarbonate connected to a system that measures the pH and the alkalinity. For sure, the reaction of the VFA with the carbonate, or bicarbonate, immediately neutralizes the pH when this latter tends to drop, though some drawbacks must be accounted for:

1. The reaction of the carbonate or bicarbonate with the acetic acid produces sodium acetate (which is precisely what the archaea need), but the alkalinity also reacts with the SH_2 (always present in biogas plant running on silage and manure). Both reactions will immediately free CO_2, hence the percentage of methane in the biogas will drop.
2. The sodium remains in the digestate at the end of the process. The most common use of digestate is as fertilizer; however, the accumulation of sodium in the ground could have the opposite effect: the salinization and consequent aridity of the soil. Beyond a certain concentration, sodium is also an inhibitor of the archaea, so the overall SMA (specific methanogenic activity) may be hampered.
3. The pH is usually measured continuously with a simple electrode, but the electrode requires an automatic cleaning system to avoid

fouling, which is not a simple task! The measured pH values may vary depending on the electrode's position within the digester. On the other hand, alkalinity can be measured only with titration or spectrophotometric methods, which are challenging to implement for continuous automatic measurement.

4 The method of the double measure, pH and alkalinity, is highly reductive and of a "linear" conception. Its validity is doubtful, despite its widespread use in Central Europe and in other countries influenced by European engineering companies, where it is known by the German acronym FOS/TAC. The installation cost of the automatic dosing devices for the alkaline solution, added to the reagents cost, may nullify the eventual benefit. When the alkalinity falls, it is more advisable to reduce the substrate's input flow and/or to increase the percentage of manure in the total feedstock. A drop in pH could be a symptom of a declining SMA and the subsequent accumulation of acetic acid, which in turn could have multiple causes.

5 A mixture of magnesium hydroxide and calcium hydroxide (spent lime), or even ash from the combustion of vegetable biomass (mainly composed of potassium hydroxide), is a good or even better additive than sodium bicarbonate and more environmentally friendly for the soil that will be fertilized with the digestate.

1.4.3 The Chemical Composition of the Substrate

This technique is another example of "management with tables". Theoretically, the content of C, H, O, and N can be directly correlated with the BMP of a given substrate. This is generally valid for pure, fully degradable substrates with a grain size of less than 1 mm. Unfortunately, the elementary chemical composition of biomass cannot be directly correlated with the speed at which methane is produced, nor does it provide any indication of how degradable the biomass will be. Several other factors will influence the BMP, such as, the composition and the metabolic activity of the bacterial ecosystem in the inoculum, the grain size of the substrate, any eventual pretreatment the substrate may have undergone that alters the degradability of the more complex molecules, etc., a full list would be too long to include here.

Some laboratories active in the biogas market calculate the BMP based on the chemical composition instead of measuring it directly. The reasons, in the author's opinion, are most probably the result of marketing interests rather than of technical considerations. The analysis of the substrate's elementary composition requires specialized instruments and qualified personnel. Hence, it remains an almost impossible task for the biogas plant manager, ensuring thus the continuous earnings to the laboratories who will benefit of "captive customers" as long as the biogas plant is active. In practice, since the BMP is calculated based on chemical analysis and corresponds to the theoretical maximum, the results are unrealistic. The biogas plant manager will be informed about high methane yields from a specific industrial substrate or a particular energy crop cultivar, in order to persuade him to purchase the substrate or seeds.

When formulating the diet of the digester based on the chemical analysis of the substrate and theoretical formulas, the results will be unreliable. As can be seen in Figure 1.10, even when digesting a pure and fully degradable substrate like starch, the real methane yield can be much different from the theoretical BMP value. As a rule, it is advisable to distrust BMP tables based on the chemical composition of silage, fodder, or other substrates. In some rare cases, the activity of the digester's bacterial ecosystem could be enough to reach the theoretical BMP under laboratory conditions, but this does not guarantee the complete degradation of the substrate under real plant operational conditions. The explanation is straightforward: recalling the Gompertz equation, as explained in Section 1.3 and illustrated in Figure 1.6, it becomes evident that the BMP represents only the coefficient *a* in the equation. The BMP tables supplied by animal fodder producers and by producers of "special cultivars for energy crops" never show the coefficients *b* and *c* of the Gompertz equation, for the simple reason that the chemical analysis method cannot calculate them.

For example, let us examine a real case illustrating the unpleasant consequences of treating table values as absolute truths. In Italy, a biomass sales representative and his client, the owner of a biogas plant, both read in a specialized agricultural magazine that fatty matters yield particularly high BMP values. Knowing that the byproducts from the oil industry always contain a certain amount of residual fats, they superficially concluded that the said type of residual biomass should be optimum for AD. In fact, the magazine in question did not include any complementary information that could have helped in defining a strategy for employing the said oily feedstock in the biogas plant. Indeed, the article defined neither the necessary time for the total conversion of fats into methane, nor the maximum gas flow that could be attainable, nor the required time to reach the said maximum flow. In other words: only the coefficient *a* of the Gompertz equation was known, but the coefficients *b* and *c* were not. Apart from being challenging to degrade and sometimes even being toxic for some bacterial ecosystems, in practice, fatty substances always require 50–60 days for their complete digestion. Hence, in spite of the high theoretical BMP resulting from chemical analysis, fatty substrates will never be able to produce relevant quantities of methane in biogas plants characterized by HRT shorter than 45 days. In our case study, the result was catastrophic: the bacterial ecosystem was not specific enough to digest fats, resulting in a production block for several days at the biogas plant.

1.4.4 The VFA Profile

This technique is theoretically accurate, but in practice, it is difficult, expensive, and too slow to apply. If we could know at any time the concentration of acetic, butyric, and propionic acids (some authors include valeric acid too), we could (based on tables from laboratory tests!) infer what the state of the AD process is. The practical problems of analyzing the profile of the VFA are multiple:

1. The chemical analysis methods (titration, fractioned distillation) of very heterogeneous volatile fatty acid mixtures characterized by low concentrations (in the order of a few mL VFA/L of slurry) are not very selective, just because the VFA are chemically very similar.
2. Chromatographic methods (gas and liquid) are quite suitable because they are selective, but require expensive instruments and the intervention of technically qualified operators. Indeed, a biogas plant manager who must simultaneously oversee farming activities, the biogas power plant, feedstock supplies, and other tasks cannot afford to purchase a chromatographer and analyze the sludge daily.
3. To be helpful for the optimization of the biological process, the determination of the VFA profile should be carried on in real time, and the information thus obtained should be the input of some "closed loop" control system, capable of controlling with nonlinear logic the stirring, the feedstock input, the pumps, the power plant, etc. Unfortunately, no sensor exists that is capable of continuously and selectively measuring the VFA concentration profile.

1.4.5 Monitoring the Biogas' Flow and Composition

Both the instantaneous gas flow and the biogas composition depend on a combination of three factors: the substrate's chemical composition, the inoculum's bacterial activity (depending in turn on external factors like temperature, pH, concentration of inhibitors like ammonia, etc.), and the ratio between the living mass of the bacterial ecosystem and the mass of feedstock to be digested. The reproduction and metabolic rates of the different species of bacteria and archaea composing the inoculum ecosystem are diametrically opposite. Under mesophilic conditions, archaea reproduce themselves more slowly than bacteria, while under thermophilic conditions, the biodiversity of fermentative bacteria is smaller. For these reasons, the proportions of hydrogen, carbon dioxide, and methane in biogas depend on the state of the digester. In this context, *state* means "a set of several biological variables" (respiratory rate, reproductive rate, concentration of the different nutrients, etc.). It could be possible to better control the AD process by establishing in the laboratory the correlation between such factors like pH, alkalinity, BMP, OLR, VFA concentration profile, concentration of dissolved H_2, and the flow and composition of biogas. If monitoring all those parameters in real time was possible, it could be then possible to employ *several tables* linked through digital controllers programmed with *fuzzy logic* (a special technology of control that, unlike digital logics that only admit 1 or 0, white or black, true or false, admits intermediate values like "partially true", "light gray", etc.). Unfortunately, although the chemical composition of biogas is relatively easy to monitor in real time, analogous to pH and the other parameters already examined, it is not a reliable single criterion for managing the biogas plant, as finding a correlation with the state of the microbial ecosystem in the digester is difficult.

1.4.6 Monitoring the ORP

The measurement of the ORP is easily performed using electric probes, similar to those used for measuring pH. As in the case of the latter, the ORP measures the *intensity* of the reactions rather than the concentration of the reagents. Since a (theoretical!) correlation between the production of acetic, propionic, and butyric acid with the ORP exists, by measuring the latter it could be possible to know the stability degree of the AD process. It is assumed that values in the range of –220 to –290 mV are good indicators of the "good health" of the process (equilibrium between the production of acetic acid and the intermediate products, propionic and butyric). ORP values within –280 and –350 mV favor the production of propionic acid and hence indicate the beginning of the inhibition of methanogenesis. On the other hand, ORP values in the range of 0 to –150 mV indicate that propionic acid is being produced, regardless of pH and alkalinity, thereby minimizing the production of acetic acid and, consequently, methane. If O_2 is bubbled in the digestion broth, so that the ORP can rise from –230 to –180 mV, the levels of SH_2 will drastically drop, but at the same time, the methanogenesis may be partially inhibited. Consequently, the ORP should be maintained stably in a restricted range, within –230 and –220 mV, a task that is not at all easy from a technical point of view.

The main disadvantage of the ORP as a control parameter for the AD process is its limited reliability due to the instability of continuous measurement, as the sensors become dirty and provide inaccurate signals. Another technical problem is to define where to place the sensor: in the laboratory it is easy to put the ORP electrode in the center of the reactor, but in the case of a digester having several thousands of cubic meters, the value measured in a given point may be different from the one measured in another point. Furthermore, ORP sensors usually have slow response times when working in anaerobic conditions—some probes may require from 8 to 24 h to stabilize their readings, a time which is long enough for undesired ORP variations to happen, which would not be detectable. Finally, ORP probes immersed in chemically aggressive environments, like digestion broth, require frequent calibrations.

The correlation between ORP and VFA concentration is not very good, as demonstrated by the experiments of Lee Sung (2008), shown in Figure 1.14.

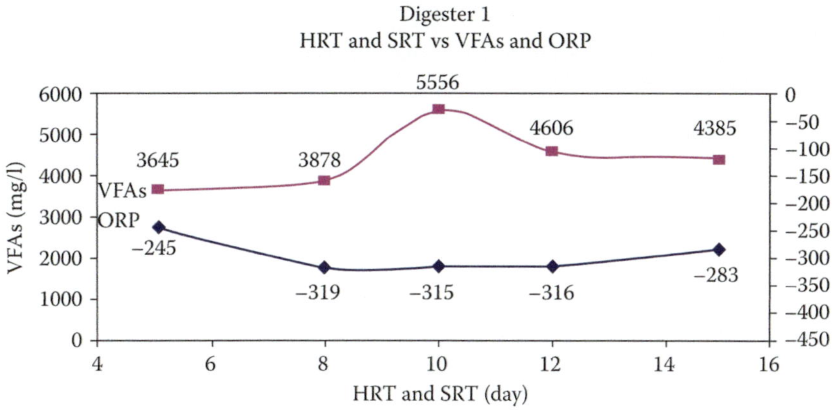

FIGURE 1.14
Concentration of VFA at different HRT, compared to the ORP values, according to Lee Sung (2008). (Courtesy of the University of Canterbury, Christchurch, New Zealand.)

1.4.7 Monitoring the Electric Conductivity

The electric conductivity is a parameter that can be easily checked with a low-cost instrument called a conductivity meter. The method's validity will be discussed in detail in Chapter 6, Section 6.2.1. In the context of the present chapter, we can state that it is the umpteenth table of values that allegedly allows a straightforward (i.e., linear) control of the biogas plant. The general idea is that the efficiency of the AD is linearly proportional to the electrical conductivity of the inoculum, as shown in Table 1.8.

The argument against such a method is that microorganisms have a high capacity to adapt to hostile environments, such as high-salinity sludge, a fact experimentally verified by the same researchers in a real-scale biogas plant. Hence, Table 1.8 should be considered merely as an indicative guideline. The author has successfully tested marine biomasses and *mozzarella* cheese whey with high salinity (approximately 40 mS/cm), observing a certain degree of inhibition compared to samples with lower salinity, but without noticing the linear and steep reduction of the BMP found by the cited researchers. It must be pointed out that a digestion process can be compromised in spite of the inoculum's salinity being at acceptable levels, and hence, measuring the conductivity does not provide a reliable diagnostic in case of problems.

The main advantages of the electric conductivity method are its low cost (<100 € for a pocket conductivity meter), its speed, and its ease of performing the measurement. The significant disadvantage is its lack of selectivity: one cannot determine whether a high conductivity value encountered in a biogas plant is due to a high concentration of salts or ammonium ions. Furthermore, the electric conductivity rises proportionally to the temperature; hence, Table 1.8 may be valid for mesophilic digesters, but it cannot be extrapolated to thermophilic plants.

TABLE 1.8

Correlation between Electric Conductivity and Net Methane Yield at 38°C

Electric Conductivity [mS/cm]	Net Methane Yield [%]
<12	100
20	85
30	65
40	32
>50	0

Source: Derived from the Curve Published by Garuti, M., et al., Conducibilità elettrica, utile monitorarla nel digestato, L'Informatore Agrario, 40, 2014.

With the limitations of the method in mind, it may be helpful to check the electric conductivity at least once a week in biogas plants fed mainly with chicken dung, or in those plants where the solid feedstock is diluted with the liquid fraction of the digestate. In the latter case, the increase in salt concentration over time is unavoidable; hence, it is advisable to use only water (preferably rainwater) to dilute the feedstock. Another reason for salinization may be the substrate itself: biomass cultivated on high-salinity soils will certainly contain more salts (ash) that will dissolve and remain in the sludge once the organic matter is degraded. Another case where checking the electric conductivity may be helpful is that of plants where caustic soda, sodium carbonate, or bicarbonate is systematically employed to control alkalinity. According to the quoted authors, the best solution to add alkalinity to the sludge without loading it with much sodium (i.e., keeping the conductivity within acceptable levels without consuming water for the dilution) consists in replacing the liquid fraction of the digestate with swine or bovine stall effluents. Nevertheless, in the latter case the concentration of ammonium ions will increase, with potentially toxic effects on the process.

1.5 The Outsourced Biological Management Service: Limitations of the Traditional Techniques

Farmers typically consult veterinarians and agronomists for the formulation of their animals' diets. So, the idea of hiring an external consultant, or paying the constructor of the plant for managing "the bacterial farm", is culturally acceptable for them. The drawback is that such *outsourced management* is more expensive and implicitly excludes the permanent surveillance of the biological process. To remain competitive, laboratories that offer biological management services limit their intervention to one weekly visit, or even a visit every two weeks, taking some samples that are analyzed according to precise routines (in general, in a laboratory located hundreds of kilometers away from the biogas plant). The result is that the biogas plant manager receives a report with the visit's outcomes several days later, when such information is no longer helpful. Quite often, the biogas plant manager does not know what to do with a report that, in general, does not contain practical indications about what to do based on the laboratory's results, and hence, he just archives the report. In other cases, the report only indicates that the biogas plant manager should add an integrator, the so-called "magical powders", punctually supplied by the plant constructor or by the laboratory itself. As an example, Figure 1.15 shows a laboratory test report, in which for privacy reasons the customer's and supplier's names were omitted.

In this case, all the tests performed by the external laboratory could have been easily carried out on site by the biogas plant manager, as they require cheap and straightforward instruments (apart from the FOS/TAC, that is indeed easy to use but requires a titrator costing roughly € 8,000 as of July 2025). The fundamental advantage of performing the tests "at home" is having the results immediately available to act immediately if necessary. Hence, it does not make much sense to pay an external laboratory for getting delayed information. In the example above, the usefulness of the results provided by the laboratory is also questionable. As discussed before, the pH serves for almost nothing. The percentages of DM, ash, and moisture (moisture is just 100% – DM%, hence including it adds no information to the report), are not very useful without a context. ORP and electrical conductivity, like pH, are of little to no use in optimizing the overall efficiency of the biogas plant, although some laboratories and plant constructors overemphasize their importance. The laboratory of this example provides some more information than others do, since it indicates how the measured values compare to average indicative values, but does not provide any proper indication on what to do to reach the recommended values.

Figure 1.16 shows the results, for the same plant and approximately at the same time, of tests carried out by another competing laboratory. In this

Analytical Profile			Minimum Average Maximum
Dry matter @ 105 °C	5,11	%	−2.55
Moisture	94,89	%	2.55
pH	7,86		0.19
Electrical conductivity	15,64	mS/cm	−1.46
Redox potential	−346,10	mV	7.50
FOS/TAC	0,27		−0.06
Ash	24,06	% D.M.	2.66
Organic matter	75,94	% D.M.	−2.66
Organic nitrogen	1801,32	mg/kg w.w.	−756.41
Ammonia nitrogen (N–NH4)	1953,56	mg/kg w.w.	−348.43

FIGURE 1.15
Example of a report about chemical analysis performed on a sample taken from a digester.

Analisi

COA n°	[redacted]	Ordine n°	[redacted]	Data	14.02.2013

Sample no.	[redacted]
Date of receipt	13.02.2013
Designation	Bioreattore
Sample type	Fermenting samples
Sampling date	07.02.2013
Sampling by	Customer
Sample amount	ca. 100 ml
Sample container	Container in PE with stabilizer
Number of containers	1
Start of analysis	13.02.2013
End of analysis	14.02.2013

Physical tests

Campione n°			13-018478-01
Sigla Cliente del Campione			Bioreattore
Dry matter	% w/w	OS	6.8
Organic dry matter (550°C)	%	OS	5.1
pH-value		W/E	7.5
Loss on ignition (550°C)	% w/w	DM	74.3

COA n°	[redacted]	Ordine n°	[redacted]	Data	14.02.2013

Short-chain aliphatic acids

Campione n°			13-018478-01
Sigla Cliente del Campione			Bioreattore
Acetic acid	mg/kg	OS	500
Propionic acid	mg/kg	OS	140
n-Butyric acid	mg/kg	OS	170
Isobutyric acid	mg/kg	OS	<50
Valerianic acid	mg/kg	OS	<50
Isovaleric acid	mg/kg	OS	<50
Caproic acid	mg/kg	OS	<50
Acetic acid equivalent	mg/kg	OS	729
VOA (titr.)	mg/kg	OS	3.900
Buffer capacity	mg/kg	OS	13.000
VOA (titr.) / TAC		OS	0.31

Nutritional values (total contents)

Campione n°			13-018478-01
Sigla Cliente del Campione			Bioreattore
Ammonium (NH4)	kg/t	OS	2.79
Ammonium-nitrogen (NH4-N)	kg/t	OS	2.2

FIGURE 1.16
Another example of laboratory report on samples taken from a digester.

case, the tests are more comprehensive and include some parameters that are useful for correct management, such as the profile of the VFA. However, the laboratory's report does not include any comments or recommendations on what the biogas plant manager should do. Once again, the shown values were measured one week after taking the sample and sent to the customer the following day, which was too late. Physical tests could have been spared, as the biogas plant manager can easily perform them in situ, as will be shown in the forthcoming chapters. They are useful only if the results are evaluated and actions are taken immediately, not one week later.

1.5.1 Conclusions

Hiring external consultants is not a wrong idea in itself, but it is necessary to choose the more professional ones. Beware of those that, apart from performing the tests, provide boosters and similar products (since their impartiality is dubious). Last but not least, it is essential that the biogas plant manager receives the test results within 24 hours of the sampling, including indications on what to do based on the measured values. External consultants or laboratories are indeed necessary to perform those tests that the biogas plant manager cannot easily carry out in the field (for instance, the profile of the VFA). The biogas plant manager should perform *in situ* the simpler tests and keep a record of the measured values to identify long-term trends.

1.6 The Automatic Titrator: Myths and Legends

From a formal perspective, the use of the automatic titrator should be included among the various techniques discussed in Section 1.4. The author considered it worth dedicating a separate section to this argument, as in some countries—mainly in Central and Southern Europe—the usefulness of the said technique has been overstated. The reason for the almost religious belief in the validity of the titration method has more to do with marketing than scientific evidence. The blind and automatic application of a simple titration method (a common practice in Chemistry) is so banal and "linear" that it easily captures the fondness of the biogas plant managers and the trust of the investors and banks that finance biogas plant projects. The current belief that a titrator is reliable, just because it is automatic and hence "foolproof", leads to the misconception that employing it as a primary diagnostic instrument will result in a reliable operation of the biogas plant, according to a linear logic based on tables. Let us begin our analysis of this technique, known as FOS/TAC or VFA/TA, by examining some theoretical concepts. FOS stands for *Flüchtige Organische Säuren*, which means volatile fatty acids in German. The final products of the organic matter's degradation are acetic, propionic, and butyric acids, called *volatile* because of their low molecular weight. They evaporate easily and cause the typical smell of rancidity. We already know that the precursor of methane is acetic acid (or, more properly, its salts: sodium, potassium, and calcium acetate). Propionic and butyric acids, if present under specific concentrations, are converted into acetic acid, and the methanogenesis process continues undisturbed. If the digester is overfed or if, for any reason, the intermediate stages of fermenting "heavy" biomasses (rich in proteins and fats) become blocked, then butyric and propionic acids begin to accumulate, consuming all the alkalinity and ultimately inhibiting the archaea by lowering the pH. On the other hand, TAC stands for *Totales Anorganisches Carbonat* (total inorganic carbonates, aka total alkalinity). In practice, the presence of carbonates produces an effect that resists pH changes, known as "buffer capacity" or simply "alkalinity". Alkalinity is relatively easy to measure with automatic instruments called titrators. The titrator can be employed to measure the *total quantity* of VFA, too, which is usually expressed *as if* they were *pure* acetic acid. We highlight "total quantity" and "as if", because what is actually present in the fermentation broth is a mixture of propionic, butyric, and acetic acids, and the exact percentage of each is unknown. So the method has a logical flaw because it is based on a false hypothesis. According to the advocates of this technique, the VFA/TA ratio indicates the health state of the AD process. Below a specific value (i.e., low concentration of VFA), the bacteria are undernourished, and the digester should be fed with more substrate. Over a particular value, the quantity of accumulated VFA risks consuming all the buffer capacity of the carbonates, which means that either the plant is overfed, or that the population of fermentative bacteria is about to overwhelm that of the archaea. In other words, it means that the latter are not able to consume all the available acetic acid, or that the penultimate link of the trophic chain, the conversion of butyric and propionic acids into acetic acid, is missing. Returning to the analogy with the instrument panel of a car: imagine reading the indication of the speedometer and the tachometer and then calculating the quotient. Theoretically, the driver would understand if the motor is running at the correct conditions for optimal fuel economy by comparing the said quotient with a table defined by the car manufacturer. In practice, such an approach is too reductive (once more, an example of the engineers' will to reduce complex phenomena to tables and linear equations!). Let us examine the weaknesses of the VFA/TA method. Table 1.9 shows the recommended values for the VFA/TA ratio.

From the explanations given, it is evident that the use of this technique can be included among the management "by tables". The FOS/TAC method is based on tests performed by some researchers under certain conditions *that do not necessarily correspond to those of our plant*. Then, somebody has redacted a table that indicates adding more or less feedstock to the digester, assuming the feedstock used is the same as the benchmark for producing the table, and without any relationship to the overall conversion efficiency into methane of the entire process. In our analogy with a car's instrument panel, the ratio between the readings of the tachometer and the speedometer could be tabulated as an indication of the optimum set point for the motor, when the vehicle runs on a flat road, in fifth gear, and the motor fed with a well-defined fuel. The table would be useless if the vehicle were climbing a slope, in a different gear, or the fuel tank contained a different type of fuel. To employ such a logic with success, a table for each possible driving condition would be required. Although nobody would ever fill the tank of his car with a fuel different from that established by the car's manufacturer, the same does not apply to anaerobic digesters, which by their own nature should digest any organic feedstock or mixtures of different biomasses, depending on their availability and economic factors. We can then conclude that a table redacted in the way described by the titrator's manufacturer—based on the operation of a single CSRT digester fed with bovine manure and corn silage—could be valid for the management of a similar plant with the same substrates. Such is the case of 98% of the agricultural biogas plants in Central Europe and most of the plants built in Italy between 2004 and 2010, but it need not necessarily be applied to a plant running on municipal wastewater sludge, or installed in other countries where corn is not the primary feedstock for AD. Even in the case of a biogas plant fed with corn silage and manure, we have observed that the degradability and, consequently, the methane yield of the former are highly variable, and the latter's features are variable as well. The BMP of manure varies depending on the age of the animals, the quantity and quality of the fodder, as well as many other factors that would be too numerous to list here. Furthermore, if the plant had two or more digesters connected in series, or if it was of the *plug flow* type (the so-called "sausage-type digesters"), it would be necessary to produce an independent FOS/TAC table for each case. If the plant's diet changes from silage to other substrates, for sure, the blind application of Table 1.9 would be, to say the least, risky.

A second problem with the VFA/TA method is its absolute lack of selectivity. Remember that the VFA value is measured *as if* it were 100% acetic acid.

TABLE 1.9

The Original FOS/TAC Table According to Lossie and Pütz (2008)

FOS/TAC Ratio	Indication	Actions to be Taken
>0.6	Excessive organic load	Stop feeding the digester
0.5–0.6	High organic load	Reduce feedstock input
0.4–0.5	The digester is at the limit	Monitor the digester carefully
0.3–0.4	Ideal conditions for the production of biogas	Keep feedstock input constant
0.2–0.3	Insufficient organic load	Increase gradually the feedstock input
<0.2	Extremely low organic load	Increase quickly the feedstock input

In practice, if only propionic and butyric acids, together with a certain alkalinity, were present, the VFA/TA ratio could be good or acceptable according to the table, but the digester would already be under biological collapse. The *said condition is not a theoretical speculation; it is relatively frequent in plants fed with silage, and the author has found it on some occasions*. The case of the collapsed digester described earlier can be explained with an analogy: imagine a clinical test that has been developed by statistically measuring the ratio between cholesterol and glycemia in a population of vegetarians and athletes, assumed to be the paradigm of "healthy" people. If the ratio falls within a specific range, the test diagnoses the person as "healthy". Now, imagine a sedentary person who is a lover of salami and sweets. It is not difficult to foresee that in such a case, the ratio between cholesterol and glycemia can easily fall within the range of values assumed as "healthy", since the test considers only the ratio and not the individual values; nevertheless, the patient in question would risk both a heart attack and diabetes.

Contrary to the "folklore" of the biogas industry in many European countries, the VFA/TA is neither a modern method nor a German invention. According to Prof. D. Bolzonella, the method was first published by Ripley et al. (1986), and its original name was "partial to total alkalinity ratio". The difference between Ripley's method and the variant called FOS/TAC by the German biogas industry is only in the final pH values assumed as the titration limits.

Why is then the FOS/TAC method so diffused in Central-South Europe, while in other countries like Sweden, England, or France it is less popular? It is not by chance that the FOS/TAC method is diffused in countries where biogas plants are subsidized, and hence, many investors have turned to this industry in search of stable profits. Since banks and venture capital firms demand to be covered by any possible guarantee, keeping a register of the VFA/TA values measured at regular intervals serves to keep them satisfied. Perhaps because pushing a button, reading a value in an instrument's display, checking a table, and finding that everything is OK sounds elementary and "linear". In most cases, the process may be running correctly, but if problems are lurking, the method does not pinpoint them. It must be noted that a biogas plant fed with cow manure and silage is unlikely to suffer biological collapse; therefore, taking a VFA/TA reading once a week or not taking it at all makes no difference in improving digestion efficiency. Furthermore, even the daily measure of the VFA/TA does not prevent the biological collapse that is always possible when a feedstock containing antinutrients, like almost all cereals, is the only substrate fed to the digester. Again, such a condition is frequent in Central Europe, Italy, and Romania, where cereal silage used to be (and to a lesser extent still is) the primary feedstock for biogas plants; however, it is undoubtedly not the case in the rest of the world. Furthermore, the use of a titrator as the sole instrument for the biogas plant's control makes it impossible to diagnose in advance

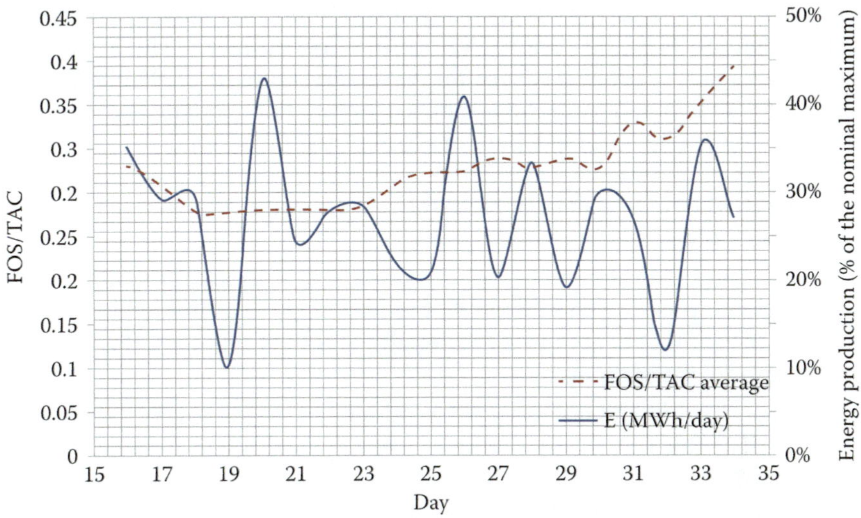

FIGURE 1.17
Comparison of the energy produced by the biogas plant and the VFA/TA parameter, measured with a FOS/TAC titrator.

the formation of hydraulic short circuits (a problem described in detail in Section 4.1.3, one of the sneakiest problems always lurking in the biogas plant).

As a demonstration of the limitations of the VFA/TA method, we will present a case study of a biogas plant in Northern Italy, with an electrical power of 1 MW, fed exclusively with corn and triticale silage, and (in small proportion) bovine manure. Figure 1.17 shows that from day 18 to 23, the VFA/TA ratio was kept at 0.25 (i.e., very close to the ideal value according to the advocates of this method). However, energy production is unstable, shifting from its maximum to its minimum peak during the period under examination. Such anomalous behavior can be explained because, at that moment, the total amount of the VFA was almost entirely butyric and propionic acids, and the methanogenic phase of the AD process was already inhibited. Observe that, according to the criterion established by the manufacturer of the FOS/TAC titrator, already shown in Table 1.9, a value equal to 0.25 means that the plant should receive its nominal daily ration (about 50 tons/day of silage). Nevertheless, in our case study, a reduced ration (approximately 12 tons/day of silage) fed from day 23 was already sufficient to cause the FOS/TAC value to grow almost exponentially, while the energy production continued to follow an erratic pattern. In this case, the FOS/TAC test was only useful to confirm what we already knew—that the plant had biologically collapsed. It didn't provide any clue on how to solve the problem.

To use the FOS/TAC titrator correctly, it is essential to follow specific practical guidelines. First, be sure to perform the test correctly. The author has found that many people measure the necessary amount of distilled water with a glass that lacks graduations, or use the automatic pipette incorrectly. It will hence be useful to refresh here the correct test procedure:

1 First, filter with a sieve or cloth the sample under analysis, to eliminate fibers and other macroscopic solids.
2 Sample the exact amount of slurry to test with an automatic micropipette, pressing the piston to the first stop, then immersing the tip in the sample to about half its length, and then releasing the piston to aspirate the sample.
3 Then load in the test glass 50 g (50 mL) of distilled water (weighted with a scale or measured with a graduated cylinder). Add the 5 mL of filtered slurry with the pipette, this time pushing its piston to the end so that the sample tip remains empty.
4 At this point, launch the automatic titration.

The values obtained with the titrator are three: the FOS parameter (VFA), the TAC parameter (TA), and the ratio FOS/TAC (VFA/TA). All three parameters must be checked as a whole set of information, and not only the nondimensional ratio FOS/TAC. For this scope, Table 1.10 is more useful than Table 1.9.

To obtain utility from the FOS/TAC titrator, the test should be performed at least once a day, always at the same time, and the measured values should be monitored in conjunction with the energy productivity of the plant, for instance, using a spreadsheet and corresponding graphic function. If some drift of the values becomes evident (productivity drops, increase in the FOS, and/or progressive decline of the TAC parameters), then complementary tests will be necessary to detect any problems and their causes. The complementary tests mentioned above (SMA test, hydrolytic activity test, analysis of the VFA profile, analysis of the oligoelements present in the sludge, analysis of ammonia and total N, measurement of electric conductivity, etc.) will be

TABLE 1.10

Reference Values of FOS, TAC, and pH in Agricultural Anaerobic Digesters, According to Daniel and Baumgartner

	FOS [mg/L]	TAC [mg/L]	FOS/TAC	pH
Normal operation	2.100	9.000	0.23	7.8
Start of acidification	3.300	8.300	0.40	7.6
Acidified reactor	4.800	6.800	0.71	7.0
Biological collapse	7.300	3.800	1.92	5.6

explained in detail in Chapters 3 and 4. In any case, the author's advice is not to rely too much on this method, since the quantity of plants that operate at full efficiency with FOS/TAC values well out of the prescribed range is probably bigger than those that comply with the literature's recommendation. The flaw in the literature is that everyone is publishing cases in which the FOS/TAC falls within the canonical range, and nobody is doing the opposite, perhaps fearing rejection of their paper or being accused of negligence. In the author's experience, confirmed by dozens of other professionals, there's no direct correlation between the FOS/TAC value and the efficiency or productivity of the digester.

Bibliography

Andrews, J. F. and Graef, S. P. *Dynamic Modeling and Simulation of the Anaerobic Digestion Process*. Anaerobic Biological Treatment Processes, Advances in Chemistry Series, American Chemical Society, Washington, D.C, 1971.

Appel, L., Baeyens, J., Degrève, J., and Dewil, R., Principles and potential of the anaerobic digestion of waste-activated sludge, *Progress in Energy and Combustion Science*, *34(6)*, 755–781, 2008.

Blanc, F.C. and Molof, A.H. Electrode potential monitoring and electrolytic control in anaerobic digestion, *Water Pollution Control Federation*, *45*(4), pp 655–667, 1973.

Bolzonella, D., *Monitoring and Lab Analysis—Assays*. Lecture of Jyvaskyla Summer School, 2013. www.valorgas.soton.ac.uk/Pub_docs/JyU%20SS%202013/VALORGAS_JyU_2013_Lecture%203.pdf.

Garuti, M., Soldano, M., and Fabbri, C., Conducibilità elettrica, utile monitorarla nel digestato, *L'Informatore Agrario*, *40*, 12–15, 2014.

Gerardi, M.H., *The Microbiology of Anaerobic Digesters*. Wiley, Hoboken, NJ, 2003.

Gerber, M., and Span, R., An analysis of available mathematical models for anaerobic digestion of organic substances for production of biogas, International Gas Union Research Conference, Paris, France, 2008.

Kosko, B., and Lupoli, A., *Il fuzzy-pensiero: Teoria e applicazioni della logica fuzzy, Collana Tascabili Baldini and Castoldi, I nani. Vita matematica; trad. di Agostino Lupoli*, 4a ed., Baldini & Castoldi, Milan, 2000.

Kuo, B., *Automatic Control Systems*. Prentice-Hall International Editions, Upper Saddle River, NJ, 1986.

Lee Sung, J., *Relationship between oxidation reduction potential (ORP) and volatile fatty acid (VFA) production in the acid-phase anaerobic digestion process, degree thesis*, University of Canterbury, 2008.

Lossie, U., and Pütz, P., *Targeted control of biogas plants with the help of FOS/TAC, Application note of Hach-Lange® DOC042.52.20011.Mar08*, 2008.

Mata-Alvarez, J., *Biomethanization of the Organic Fraction of Municipal Solid Wastes*. IWA Publishing, London, UK, 2002.

Ogata, K., *Modern Control Engineering*. Prentice-Hall International Editions, Upper Saddle River, NJ, 1985.

Prokopchenko, O.E., The Gompertz Sigmoid function and its derivative, The Wolfram Demonstrations Project. 2011. http://demonstrations.wolfram.com/TheGompertzSigmoidFunctionAndItsDerivative/.

Raposo, F., De la Rubia, M.A., Fernández-Cegrí, V., and Borja, R., Anaerobic digestion of solid organic substrates in batch mode: An overview relating to methane yields and experimental procedures, *Renewable and Sustainable Energy Reviews*, *16*(*1*), 861–877, 2011a.

Raposo, F., et al., Biochemical methane potential (BMP) of solid organic substrates: Evaluation of anaerobic biodegradability using data from an international interlaboratory study, *Journal of Chemical Technology and Biotechnology*, *86*(*8*), 1088–1098, 2011b.

Ripley, L. D., Crow, V. W., Ziegler, L. J., and Park, W. H., *An empirical method to estimate the ratio of volatile fatty acids to total alkalinity in anaerobic digesters.* Proceedings of the 5th Symposium on Biotechnology in Energy Production and Conservation, 198–203, 1986.

Schnürer, A., and Jarvis, A., *Microbiological Handbook for Biogas Plants*, Swedish Waste Management U2009:03, Swedish Gas Centre Report 207, 2010.

Van Lier, J., Sanz, M., and Lettinga, G., Effect of temperature in the anaerobic thermophillic conversion of granular and dispersed sludge, *Water Research*, *30*(*1*), 199–207, 1996.

Varnero Moreno, M.T., *Manual de biogás, Edition by FAO (Food and Agriculture Organization) and Government of Chile, Santiago, Chile, ISBN 978-95-306892-0*, 2011.

Violante, P., *Chimica e fertilità del suolo.* Edagricole, Bologna, Italy, 2013.

2

Overview of Laboratory Methods for the Analysis of Fermentative Processes

2.1 Basic Notions of Metrology: Accuracy and Precision or Repeatability

> I often say that when you can measure what you are speaking about, and express it in numbers, you know something about it; but when you cannot measure it, when you cannot express it in numbers, your knowledge is of a meagre and unsatisfactory kind; it may be the beginning of knowledge, but you have scarcely, in your thoughts, advanced to the stage of science, whatever the matter may be.

These words of Sir William Thomson (better known as Lord Kelvin, because of the temperature scale named in his honor), published in 1883, represent the essence of this book. Running a biogas plant requires constantly measuring a series of parameters to optimize the process, thereby increasing the amount of methane per cubic meter of digester (process optimization) and/or per ton of feedstock (substrate optimization). The question is: What does the word "to measure" mean in this context?

> To measure means to compare an unknown magnitude with another magnitude assumed as a measurement unit.

This is relatively easy when one intends to measure a well-defined physical unit, for instance, a weight, a length, or an electric current. On the contrary, it is challenging to "measure" a complex process, like anaerobic digestion, because we cannot count the amount of living bacteria of each species and their respective metabolic activities and interactions. We can only define the *state* of the entire system (microbial population and feedstock) based on a set of measurements (pH, temperature, concentration of VS, etc.) that correlate directly with the variable we desire to control or improve (instantaneous flow and gas composition).

DOI: 10.1201/9781003676201-2

We can add another truth to Lord Kelvin's words, apparently obvious but not always proving true in practice: Apart from *being able to measure*, it is also necessary to *know how to measure* correctly. Hence, before discussing some practical examples, in the next paragraphs, we are going to introduce some theoretical concepts about how to perform reliable measures.

2.1.1 Definitions of Accuracy and Precision or Repeatability

Figure 2.1 summarizes the definitions of both concepts with the example of a keen sniper who tests four different rifles or, conversely, of four snipers with various degrees of ability testing the same gun. The snipers can fail to hit the center of the target. The analogy of snipers and rifles within our context allows us to state that measurement errors can result either from an incorrect calibration of the instrument or the user's inexperience or carelessness.

A good instrument must be precise, i.e., it must measure values that are consistent in time, and it must be accurate too, i.e., it must measure values very close to the true value of the physical magnitude (case A).

An instrument that is precise but not accurate (case B) induces a systematic error, easy to compensate for, on the condition that both the amount and the sign of the uncertainty are known.

An instrument that is accurate but not precise (case C) induces random errors. Such kinds of instruments are acceptable for maintenance scopes, because the average value of the measures will be close to the true value, but will require a relatively high number of measures to obtain a representative average value.

An instrument that is neither precise nor accurate (case D) should not be employed, because the measures will be unreliable.

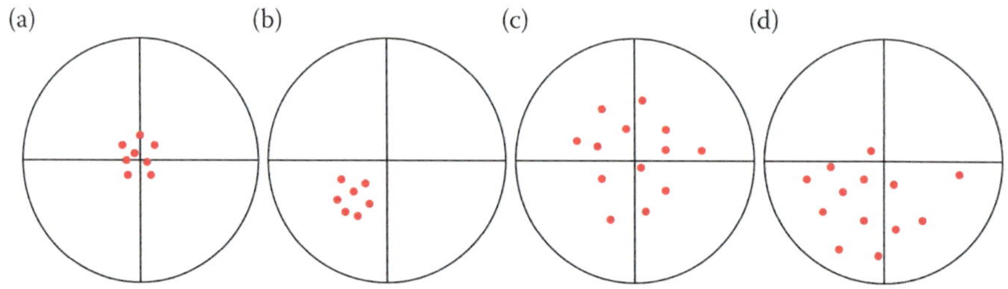

FIGURE 2.1
Graphic analogy of the concepts of accuracy and precision, or repeatability. (a) Accurate and precise. (b) Precise, but not accurate. (c) Imprecise (or disperse), but the average value may be acceptable because the dispersion is not biased in a given direction. (d) Imprecise and inaccurate. (https://en.wikipedia.org/wiki/Accuracy_and_precision.)

2.1.2 Error Propagation

There are two different possibilities for measuring a given physical magnitude:

1 We *measure it directly* with a *dedicated* instrument;
2 the unknown physical magnitude results from a *calculation obtained from one or more measurements.*

When measuring a physical magnitude directly, the error (or uncertainty) of the measurement depends directly on the instrument (and/or on the user's skill). When the unknown physical magnitude is derived from one or more measurements, the total error or uncertainty of the calculated value will be a function of the single measurements and their respective errors (usually the sum of the single errors or an overall error amplification factor). To understand how the errors propagate when a value is derived from several measurements, it is possible to employ a series of rules, which are based on theorems. Since the scope of this book is to prioritize practical experience over theory, the theorems' demonstrations will not be included here, and the error propagation will be estimated in all cases by the application of the said rules. We will start by defining the error types and their symbols, presenting a few examples that will help us to understand how to apply the error propagation rules.

Definition no. 1

$E_{(x)}$ = absolute error or uncertainty of the measured magnitude x.

If x_m is the value of a physical magnitude measured in a given way and x_v is the true value (in general unknown), the absolute error $E_{(x)}$ is defined as:

$$E_{(x)} = x_m - x_v$$

$E_{(x)}$ is hence expressed in the same unit of the measured magnitude (m, kg, V, etc.).

EXAMPLE OF ABSOLUTE ERROR

Consider a graduated cylinder with 100 mL nominal capacity, graduated with 1 mL divisions. The absolute error of a liquid volume measured with this device is ±1 mL, because the intermediate values cannot be measured with certainty. Although it is possible to observe with the naked eye if the level of the liquid is between two divisions, it is not possible to state if it is in the exact middle.

Definition no. 2

$e_{(x)}$ = relative error or uncertainty of the measured magnitude x.

The relative error $e_{(x)}$ is defined as:

$$e_{(x)} = \frac{E_{(x)}}{x_v}$$

Consequently, $e_{(x)}$ is nondimensional and will be expressed in %.

EXAMPLE OF RELATIVE ERROR

Consider the same graduated cylinder in the former example. If 20 mL of liquid is measured, the uncertainty of the measurement is equal to ±1 mL, the relative error is equal to 1/20, and hence ±5%. If, instead, 100 mL of liquid is measured, the relative error will be 1/100, i.e., ±1%.

FIRST PRACTICAL CONCLUSION

When a physical magnitude is measured with whatever instrument, it is good practice to choose the instrument's scale such that it results as close as possible to the estimated value of the magnitude. For instance, if it is necessary to measure volumes in the order of 9–12 mL, it is convenient to get a graduated cylinder with 15 mL nominal capacity, because employing bigger graduated cylinders will induce a bigger relative error.

Now that the concepts of relative and absolute errors of a measured value have been defined, the fundamental theorems of the *Theory of Errors* can be stated as the following practical rules:

1 Relative error of the product of two measured magnitudes, $x \cdot y$

$$e_{(x \cdot y)} = e_{(x)} + e_{(y)}$$

2 Relative error of the quotient of two measured magnitudes, x/y

$$e_{(x/y)} = e_{(x)} + e_{(y)}$$

3 Absolute error of the sum or the difference of two measured magnitudes, $x \pm y$

$$E_{(x \pm y)} = E_{(x)} + E_{(y)}$$

4 Relative error of the average of n measured magnitudes $x_1, x_2, \ldots, x_n = \overline{x}$

$$e_{(\overline{x})} \approx e_{(x)}$$

The former rule is valid only on condition that the single values $x_1, x_2, \ldots, x_n$ are of the same order of magnitude as $\overline{x}$, and furthermore that the value of $\overline{x}$ is as close as possible to the estimated value of the magnitude (condition of minimum relative error).

SECOND PRACTICAL CONCLUSION

From the former rules, it can be deduced that the maximum error made when calculating a value with a formula, based on physical variables measured independently, depends on the formula and the method employed, hence not only on the precision and accuracy of the instruments.

Conventions for the correct expression of measured values, or of values calculated from measures, and their errors

1 $E(x)$ is always written with a single significant figure.

2 The result of a measurement whose $E(x)$ is known is written with all the significant figures up to the same order of the error, rounded.

Let us see some practical examples of how to apply the rules and conventions.

EXAMPLE 1

Measure the dry matter (DM) of a substance, planning the measurement in such a way that, when employing a simple digital kitchen scale with good criterion, the relative error will be less than 1%

Data:
Accuracy of the scale is ±0.5 g (in general, this value is provided by the manufacturer in the user's manual or the technical sheet)

Order of magnitude of the DM (estimated) ≈ 30%

By definition:

$$\mathrm{DM}[\%] = \frac{\mathrm{DW}[\mathrm{g}]}{\mathrm{WW}[\mathrm{g}]}$$

where DW is the dry weight and WW is the wet weight.

$$e_{\mathrm{WW}} = \frac{E_{\mathrm{WW}}}{\mathrm{WW}}$$

$$e_{\mathrm{DW}} = \frac{E_{\mathrm{DW}}}{\mathrm{DW}}$$

Since the same scale will be employed for measuring both the humid and the dry matter (after drying a sample in an oven), the absolute errors of both weights are:

$$E_{\mathrm{DW}} = E_{\mathrm{WW}} = \pm 0.5\,\mathrm{g}$$

Applying the rule of the quotient of two measured values, we obtain:

$$e_{\mathrm{DM}} = e_{\mathrm{WW}} + e_{\mathrm{DW}}$$

Since the goal is to keep the error in the determination of DM smaller than ±1%, the former equation becomes:

$$0.01 = e_{\mathrm{WW}} + e_{\mathrm{DW}}$$

Since we assumed that:

$$\mathrm{DW} \approx 30\%\,\mathrm{WW}$$

we can hence write:

$$e_{\mathrm{DW}} \approx \frac{E}{0.3 \cdot \mathrm{WW}} = \frac{0.5\,\mathrm{g}}{0.3 \cdot \mathrm{WW}}$$

$$e_{\mathrm{WW}} = \frac{E}{\mathrm{WW}} = \frac{0.5\,\mathrm{g}}{\mathrm{WW}}$$

Hence:

$$0.01 = \frac{0.5\,\mathrm{g}}{0.3 \cdot \mathrm{WW}} + \frac{0.5\,\mathrm{g}}{\mathrm{WW}}$$

$$0.01 = \frac{(0.5 + 1.666)\,\mathrm{g}}{\mathrm{WW}}$$

$$\mathrm{WW} = \frac{(0.5 + 1.666)\,\mathrm{g}}{0.01} = 216\,\mathrm{g}$$

FIRST PRACTICAL CONCLUSION

It is possible to measure with good accuracy even when employing inaccurate instruments, on condition that the measurement procedure is planned with a correct criterion.

EXAMPLE 2

A student measures the DW of a digestate sample with a laboratory scale, whose error margin (defined by the manufacturer) is ±5 mg. The wet sample weighs 1 g, and the weight after drying it in the oven is 0.045 g. Estimate the error in the determination of the DM under said conditions.

By definition:

$$e_{\mathrm{DM}} = e_{\mathrm{WW}} + e_{\mathrm{DW}}$$

$$e_{\mathrm{WW}} = \frac{E}{\mathrm{WW}} = \frac{0.005\,\mathrm{g}}{1\,\mathrm{g}} = 0.5\%$$

$$e_{\mathrm{DW}} = \frac{E}{\mathrm{DW}} = \frac{0.005\,\mathrm{g}}{0.045\,\mathrm{g}} = 11.1\%$$

$e_{\mathrm{DM}} = 12\%$ (remember that, conventionally, the error must be rounded to its nearest significant figure).

SECOND PRACTICAL CONCLUSION
It is of no use having accurate (expensive) instruments if they are employed with the wrong criterion!

EXAMPLE 3

It is necessary to measure the DM of several biomasses and then calculate the average DM of the mixture, with the scope of defining the diet to feed a digester. The available instrument is a halogen moisture scale that shows the result directly as DM, with ±1% uncertainty. The manufacturer declares such accuracy for samples of at least 10 g, but the technical sheet does not state if the accuracy is constant throughout the measurement range, 0%–100%. According to the manufacturer, the uncertainty of the instrument when working in "scale mode" is 0.2%. By an international convention, the said percentage refers to the "end of scale" aka "scale limit", in this case, the maximum weight reading of the instrument, which is 100 g. This is equivalent to saying that the absolute error of the instrument in "scale mode" is 0.2 g.

Calculate the relative error of the total VS in the daily ration when employing the average instead of the single values.

N.B. Please note that, since the instrument measures the DM directly in % of the wet weight, its measurement unit is "%". Hence, "±1% uncertainty" in this context means that the *absolute error* is equal to "1%."

Values of DW% measured on 10 g samples:

Biomass A = 12%

Biomass B = 38%

Biomass C = 60%

The average is then:

$$\overline{DM} = 36.66\% \approx 37\%$$

Remember: By convention, the calculated value is rounded to the nearest significant figure. Since the instrument is unable to detect variations smaller than 1%, i.e., it cannot distinguish between a sample with 36% DW and a sample with 37% DW, it would be absurd to express the measured value with decimals. Such an error is very frequent in the literature.

CALCULATION OF THE ERRORS

Since the single values do not have the same order of magnitude of their average, the approximate rule derived for the error of the average is not valid in this case. By definition:

$$\bar{D}\bar{M} = \frac{DM_1 + DM_2 + DM_3}{3}$$

$$e_{\bar{D}\bar{M}} = \frac{E_{(DM_1+DM_2+DM_3)/3}}{(DM_1 + DM_2 + DM_3)/3}$$

Since 3 is a figure, it has no associated error, so:

$$E_{(DM_1+DM_2+DM_3)/3} = E_{(DM_1+DM_2+DM_3)} = E_{DM_1} + E_{DM_2} + E_{DM_3} = 3 \cdot E_{DM}$$

$$e_{\bar{D}\bar{M}} = \frac{3 \cdot E_{DM}}{\left(DM_1 + DM_2 + DM_3\right)/3} = \frac{0.03}{0.37} = 81\%$$

Please note that performing three separate measures and then calculating their average introduces an absolute error greater than the individual error of each measure, especially if the single values are significantly different from the average. In this case, the absolute error of the average value is three times the error of the single measure. Alternatively, we can also say that the average value has an uncertainty of 81% (meaning 81% uncertainty of the calculated 37%). Both forms are equivalent; we must only pay attention not to confuse what "percent" means in each case. Summarizing: By individually measuring the DM percentage of each biomass, we calculate an average DM of the mixture as 37%, but, in practice, the actual mixture could have any percentage of DM in the range of 34%–40%.

THIRD PRACTICAL CONCLUSION

When calculating average values of measures taken far below the maximum reading of the instrument, and if the individual readings are of a different order of magnitude than the average, the error tends to be amplified. In our example, it would have been easier and more accurate to take a 100 g sample from each biomass, mix the samples very well, then extract at least 10 g of the homogeneous mixture, and measure its DM directly. We would have obtained the desired value with just one single test, and with an absolute error margin of ±1%.

EXAMPLE 4

Suppose weighing a sample of biomass with a halogen moisture analyzer set in "scale mode". The maximum weighing capacity of the instrument is 110 g. The manufacturer states that the maximum error is ±0.02% and that the resolution is 0.005 g.

The reading of the scale is 22.385 g. Is it correct to adopt this value for further calculations?

Since the scale error is 0.02% of 110 g, then:

$$E_{(x)} = 0.0002 \times 110 \text{ g} = 0.022 \text{ g}$$

By convention, $E_{(x)}$ must be rounded to its first significant figure; hence:

$$E_{(x)} = \pm\, 0.02 \text{ g}$$

The specification of the manufacturer "resolution = 0.005 g," means that the instrument can discriminate between two samples weighing X.XX0 g and X.XX5 g. As we can see, the maximum error of the reading is bigger than the resolution (this a typical feature of low-cost moisture analyzers, because in "moisture test mode" the important feature is the *difference* between the initial and the final weight, independently of the accuracy of the absolute initial and final weights themselves). Practically, in "scale mode," we are not able to know if the real weight of the sample under examination is 22.385 g, because under the measurement conditions of the present example, the actual weight could be any value in the range of 22.365–22.405 g.

The relative error of the reading is then:

$$e_P = \frac{0.02 \text{ g}}{22.385 \text{ g}} = 0.000893 \approx \pm 0.09\%$$

The correct way to express the reading is:

$$x = 22.38 \text{ g} \pm 0.02 \text{ g}$$

FOURTH PRACTICAL CONCLUSION

The former example applies especially to digital instruments: The number of decimals shown in the instrument's display does not necessarily imply a higher accuracy of the reading. In the previous example, the displayed reading has three decimals, but only the first two are significant.

2.2 Measurement Methods Employed in the Biogas Industry

We have seen in Chapter 1 that the optimization of a biogas plant requires monitoring a series of physical and chemical magnitudes. Some of them (for instance, the profile of the volatile fatty acids [VFA]) require the services of specialized laboratories, since they possess sophisticated instruments to carry out complex analytical techniques.

On the other hand, the biogas plant manager can easily measure many other important variables *in situ*, using a few standard instruments that do not require scientific skills. The said variables provide, in real time, valuable insights about the necessary actions to keep the anaerobic digestion (AD) process fully efficient. The practical examples 1 and 2 in Section 2.1 are useful to make the key concept clear: if we know *how to* measure using the available instruments, *we can* measure correctly, and hence, we can *improve* our process. In this section, we will present an overview of the tests that the biogas plant manager can perform, the necessary instruments for each scope, and some guidelines for the selection and purchase of the most suitable instruments. The test methods and the practical application of the results obtained will be discussed in Chapters 3 and 4, respectively.

First, we can classify the *in situ* tests into two big categories: physicochemical tests and biological tests. Physicochemical tests (on biomass, slurry under digestion, or digestate) are somehow preparatory to the biological tests and altogether provide indirect information from which we can easily infer the state of the microbial ecosystem and the quality of the biomass employed as feedstock. Table 2.1 summarizes the more suitable tests to be performed *in situ* by the biogas plant manager, because of their simplicity and the importance of the information obtained with them.

2.2.1 Classical Volumetric and Barometric Methods

The most useful tests for analyzing the AD process are the biological ones. All those shown in Table 2.1 are called *respirometric tests* because they are based on measuring the quantity of methane (or biogas) produced in a given time under certain conditions, from which it is possible to deduce the necessary information about the living mass and the metabolic activity of the microbial ecosystem.

Two "classical" schools of thought exist about how to measure the volume of gas produced per unit of time:

1. *Measurement at constant (or almost constant) pressure*: This method is based on the principle that, if the pressure of a gas is kept constant, its volume will be proportional to the mass of gas produced by the AD. Those methods employing the said principle are called *volumetric*

TABLE 2.1

Useful *In Situ* Tests for Optimizing Biogas Plants

Physicochemical Tests	Measured Variable	Necessary Instrument
DM	% of total solid matter	Halogen moisture analyzer, or an oven, and a precision scale
Ash (A)	Mineral fraction of the DM (%)	Muffle oven and precision scale
Volatile solids (VS)	% of organic matter in the solids	Calculation: VS = DM-A
COD	% of organic matter (in slurry and liquid substrates)	Photometer and pre-dosed vials
N ammonia, N total	Dissolved NH_4^+ and total N (sludge and liquid substrates)	Photometer and pre-dosed vials
pH/ORP	pH/ORP	pH meter and ORP-meter
Composition of biogas	% of CO_2, and by difference, % of CH_4	Syringe, plastic tubes, and caustic soda, or a portable gas analyzer
Biological Tests	**Measured Variable**	**Necessary Instrument**
BMP RMP	CH_4 productivity of the feedstock Residual CH_4 productivity of the digestate	Batch reactors and counters of the accumulated gas volume, ideally also a data logger to minimize manual work
AD curve	Kinetics of the AD process (the parameters *a*, *b*, and *c* of the Gompertz equation)	Batch reactors and counters of the accumulated gas volume, ideally also a data logger to minimize manual work
Specific Methanogenic Activity (SMA)	Capacity and speed of degrading acetate (and/or VFAs in general)	Batch reactors and counters of the accumulated gas volume, ideally also a data logger to minimize manual work
Hydrolytic activity	Capacity and speed of degrading starch and cellulose (polysaccharides in general)	Batch reactors and counters of the accumulated gas volume, ideally also a data logger to minimize manual work
Proteolytic activity	Capacity and speed of degrading proteins	Batch reactors and counters of the accumulated gas volume, ideally also a data logger to minimize manual work
Lipolytic activity	Capacity and speed of degrading lipids (oils and fats)	Batch reactors and counters of the accumulated gas volume, ideally also a data logger to minimize manual work

because the measurement device counts the volume of gas produced (e.g., by measuring the displacement of a piston, the displacement of a water column, and any kind of mechanical counter).

2 *Measurement at constant volume:* This method is based on the principle that, if the volume of the digester is kept constant, the pressure on the walls produced by the biogas is directly proportional to its mass. Those

methods employing the said principle are called *barometric*, since the measurement device (mechanical manometer, mercury column, solid-state pressure sensor, etc.) senses the pressure of the gas produced by the AD.

A third school of thought, limited to a fraction of the academic world, proposes the gas chromatographic method. It is similar to the barometric method, but instead of measuring the increase in pressure, the instrument measures the variations in the composition of the gas. The said method was described by Hansen et al. (2004). It has an acceptable accuracy (about 2% instrumental error), but its sensitivity is low. According to the cited authors, it cannot detect variations smaller than 75 Ncm^3, which is a huge absolute error if we consider that volumetric methods can discern volumes as low as 2 mL. Furthermore, the gas chromatograph is an expensive instrument that requires a skilled operator and frequent calibration, so this method is unsuitable for industrial use.

Finally, two novel methods were proposed by Hafner et al. in 2020: the gravimetric and gas density methods. Both methods are manual, based on measuring the difference in weight of a reactor as biogas is vented. In the author's opinion, both are acceptable for didactic purposes at a university, as the required equipment is minimal, but unsuitable for the biogas plant manager because, like every manual method, they are time-consuming. Furthermore, measuring tiny differences in weight implies summing up the absolute error at each venting, so the final relative error grows with the test duration.

Now let us analyze some of the possible technical solutions, both handcrafted and commercial, together with their advantages and disadvantages.

2.2.1.1 Volumetric Methods

Figure 2.2 shows one of the most rudimentary systems possible: the displacement of a piston in a conveniently graduated cylinder. It was first adopted by the University of Hohenheim in Germany, and hence it is usually cited in the literature as the "Hohenheim system". For its construction, only a few syringes are required, which will simultaneously serve as the fermenter and measurement system, along with a thermostatic bath to maintain the desired temperature (or an electric incubator). The only advantages of this solution are its extremely low cost and the readily available materials on the market. Its most important disadvantages are the following:

- The friction coefficient of the piston is not constant; hence, it will displace jerkily, and therefore, the reading is not reliable.
- The volume of most of the syringes available in the market is usually smaller than 100 mL; hence, it is necessary to work with tiny quantities of inoculum and substrate. Significant extrapolation errors are then induced, and in most cases, the representativeness of the sample is lost.

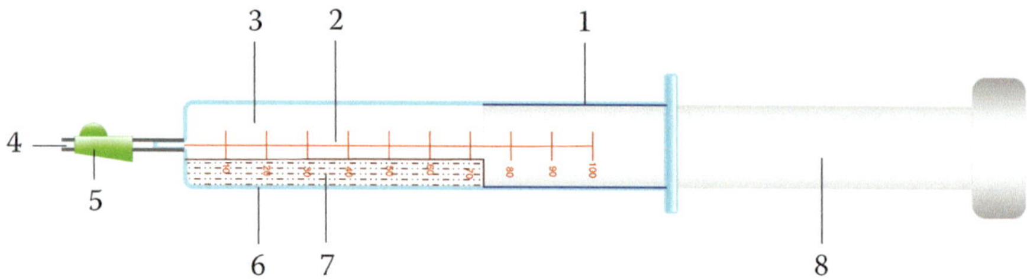

FIGURE 2.2
Volumetric measurement test using a syringe. (1) Oil or silicone grease (both for lubrication and sealing), (2) Graduations, 1 mL, (3) Volume occupied by the biogas, (4) Output valve for the gas chromatographer, (5) Valve, (6) Glass syringe, (7) Slurry under digestion, and (8) Piston. The set of syringes is placed in a rotating support, inside a thermostatic cabinet or incubator. (Drawing by the author, based on the draft of norm VDI 4630/2014.)

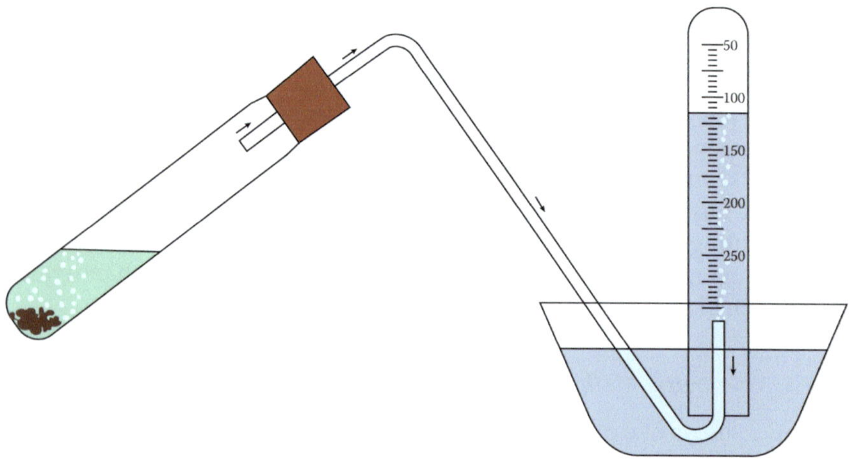

FIGURE 2.3
An elementary eudiometer. (www.wpclipart.com/science/chemistry/eudiometer.png.html.)

- The small volume of the syringe requires taking readings quite often. Should the volume of produced gas be bigger than the syringe's volume, the piston could be pushed out of the cylinder, losing all the gas and spoiling the test.
- The measurement error is quite difficult to estimate because of the piston's jerky displacement.
- It is mandatory to get a weather station to measure the room's pressure and temperature, and be able to normalize each single reading. This adds the errors of both the thermometer and the barometer to the calculation of the normalized biogas volume.
- The small diameter of the syringe's aspiration tube limits this method to liquid substrates or very diluted organic matter, for instance, wastewater or nutrient solutions prepared in the laboratory.

Figure 2.3 shows the operation scheme of another easy-to-build instrument: the eudiometer, invented by Alessandro Volta for his experiments with what he called "air that burns" (biogas released by decaying organic matter in marshes, a.k.a. marsh gas). In the eudiometer, the "piston" is a liquid column (usually water; oil is seldom employed, although it has some potential advantages over water).

Figure 2.4 shows a homemade eudiometer, and Figure 2.5 shows a commercial bank of eudiometers.

The eudiometer is an instrument invented by Alessandro Volta in 1777, conceived to measure the quantity of gas produced by chemical reactions (quote from Alessandro Volta Foundation, www.alessandrovolta.it). It can be well adapted for performing respirometric tests. The pros and cons of the eudiometer, when employed for anaerobic tests, are summarized in Table 2.2.

Another variant of the volumetric methods is the so-called "fluid displacement". Basically, it consists of two recipients: one working as an anaerobic

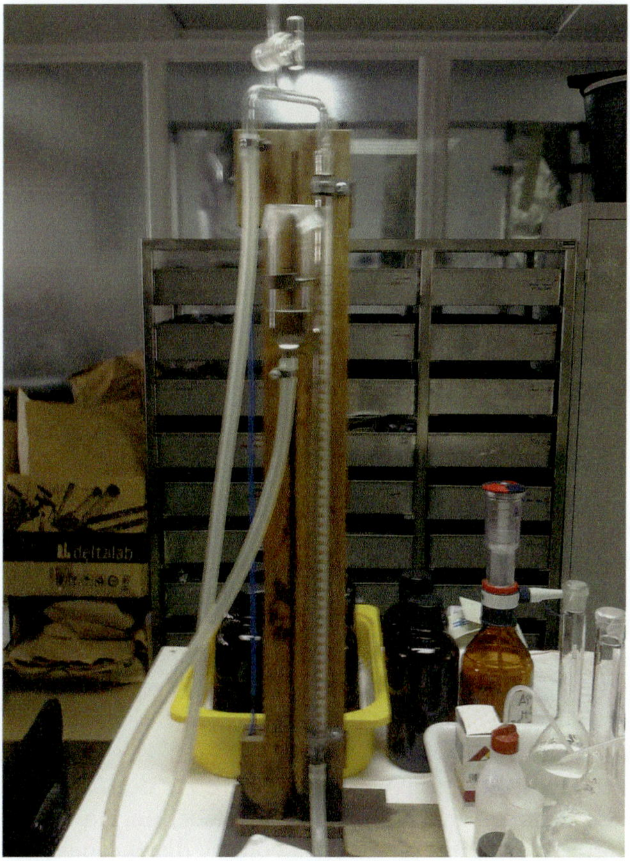

FIGURE 2.4
"Vintage" eudiometer conserved at CIEMAT of Madrid. (Photo by the author.)

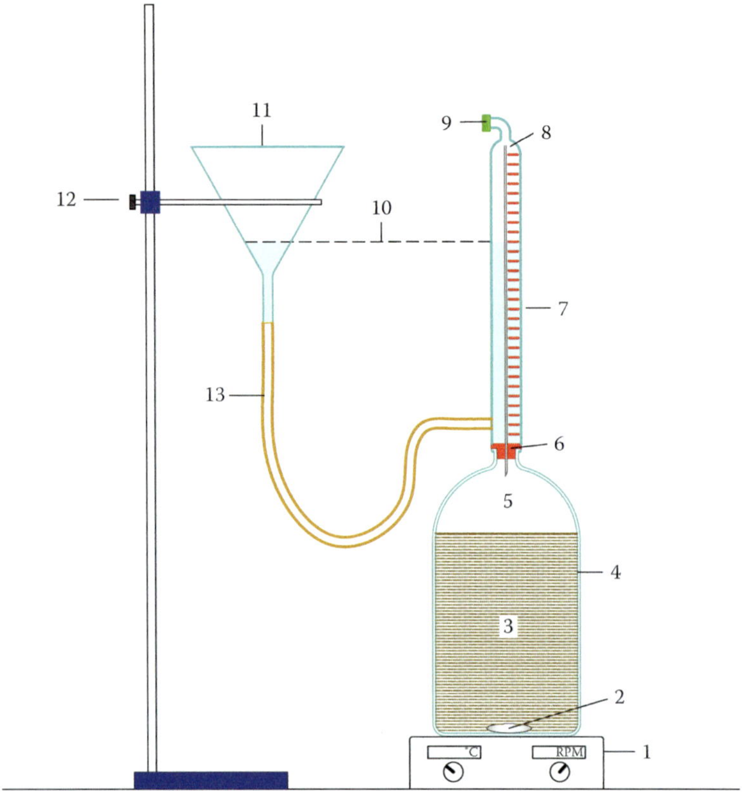

FIGURE 2.5
Eudiometer according to the DIN 38414 Standard. (1) Heated magnetic stirrer, (2) Magnetic bar, (3) Slurry under test, (4) Glass bottle, (5) Headspace, (6) Joint of the eudiometer tube (rubber, plastic, or sanded glass), (7) Graduated eudiometer tube, (8) Gas collection space, (9) Gas sampling and venting port, (10) Water level, (11) Container for collecting the displaced water, (12) Adjustable height stand, and (13) Rubber tube. (Drawing by the author.)

reactor (hence it is possible to add a suitable stirrer and a thermostatic bath to maintain a stable temperature) and another recipient containing liquid, having a siphon through which the gas will displace the liquid, which is collected in a graduated cylinder. Since the displaced volume is equal to the volume of gas, the accuracy of the measure depends on the accuracy of the graduated cylinder, on condition that, between consecutive readings, the variations of ambient T and P are negligible. According to the norm ISO DIN EN 1042, two qualities of graduated cylinders exist: class A and class B (class B has twice the error of class A). The absolute error depends on the nominal volume of the cylinder and is stated on the label on the device, as can be seen from Figure 2.6. Figure 2.7 shows a volume displacement measurement system built by one of the author's students.

TABLE 2.2

Advantages and Disadvantages of the Eudiometer for Measuring Biogas from AD

Advantages	Disadvantages
Simple operation, no moving parts	Fragility of the glass tubes.
If the liquid is a solution of NaOH in water, the instrument measures directly the CH_4 produced. If oil is employed, it can measure total biogas with minimum error caused by the solubility of CO_2 in water.	The accuracy of the instrument's reading is proportional to the height of the liquid column, but a high column implies that the operation principle is not any more "at constant pressure", and this induces errors in the normalization of the gas volume.
Parts easily available in the market.	Fabrication requires highly skilled craftsmen; medium to high cost.
Easy reading.	Only manual reading = possibility of human error + much time to dedicate to the test.
Possibility of self-construction.	Impossible to include a stirrer robust enough for agricultural slurry and biomass.
	Requires a special thermostatic bath.
	Readings subject to variations of ambient *P* and *T*; so it is necessary to have a thermometer and barometer and also take these readings to compensate the normalization errors.
	Working volume in general is small; hence, during the weekends it could remain dry and spoil the whole test.

Figure 2.8 shows the scheme of a solution, more sophisticated and expensive than the handcrafted construction in Figure 2.7, based on standard laboratory glassware. The accuracy of this second solution will not necessarily be higher than the handcrafted one: Remember that the accuracy is exclusively related to the class of the graduated cylinder measuring the volume of displaced liquid.

Compared to the eudiometer, volume displacement systems present the advantage of easier "DIY construction" (even with common materials, not necessarily laboratory glassware), the possibility of foreseeing a bigger displacement volume (hence no risk of losing the test's result during weekends), constant pressure (because the column height of the NaOH solution will stabilize the pressure), and, in general, lower cost. The disadvantages of volume displacement methods are the same already reported for the eudiometer in Table 2.2.

Finally, the family of the volumetric methods includes the volumetric gas counters. In general, these devices are based on a combination of hinged bells immersed in a fluid, paddle wheels, small turbines, or even optical systems that count the bubbles released by a small-diameter tube submerged in a liquid. Such instruments are more complex and, in general, include some electronic printed circuits and are delivered with a calibration certificate. Their cost is definitely higher than the simple systems we have described so far, but on the other hand, their operation is automatic (at least to a certain

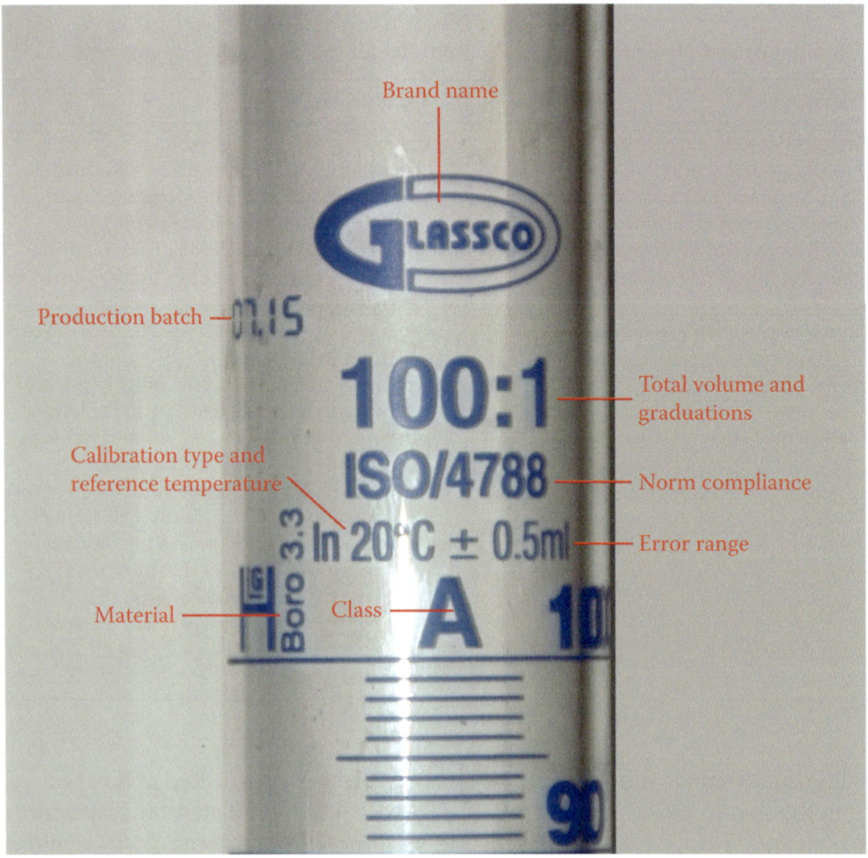

FIGURE 2.6
Example of class A cylinder labeling according to ISO DIN EN 4788. (Photo by the author.)

extent), hence they do not require the permanent attention of an operator. For example, Figure 2.9 illustrates the BPC Go instrument, whose photo is reproduced courtesy of BPC Instruments AB.

It is difficult to compile a table with the advantages and disadvantages of these types of measurement instruments, since their features vary enormously from one manufacturer to another. The model shown in Figure 2.9 is one of the most modern models currently on the market. It includes ambient pressure and temperature sensors, so its reading is already normalized in real time, interchangeable cells of 2 and 9 mL, datalogging with report download in Excel format, and Ethernet interface. Other similar instruments lack these unique features, so the readings and their normalizations must be performed manually. The accuracy varies significantly, ranging from ±1% to ±5% depending on the brand and model. The resolution (minimum volume of gas that the instrument can measure) generally lies in the range of 2 to 20

FIGURE 2.7
Water displacement measurement system built by Mr. Gianluca Bergamaschi, employed to optimize the digester of Azienda Agricola Valsesia s.n.c. (Photo kindly provided by its owner.)

mL, with some models being unable to detect volumes smaller than 100 mL. The only feature that all analogue instruments have in common is the manual reading (either on a digital display or on an analogue quadrant). Hence, the curves of biogas production obtained with these kinds of counters are usually polygonal. To obtain smoother curves with higher resolution, it is necessary to take at least four readings per day. Additionally, reading the room's P and T (if the counter lacks built-in automatic normalization) will also be required. Consequently, a laboratory thermometer and barometer will be necessary, which adds the respective errors to the normalized volume. In some models, the counters have a standard analogue signal output (4–20 mA or 0–10 V), proportional to the flow of measured gas. In such case, it is then possible to connect the said analogue output to a standard data logger system, so as to monitor the gas production continuously and to obtain data in numeric format (Figure 2.10).

2.2.1.2 Barometric Methods

Most elementary barometric systems consist of a bottle with a cap tightly locked to the bottle's neck using a clamp (somewhat like a champagne bottle).

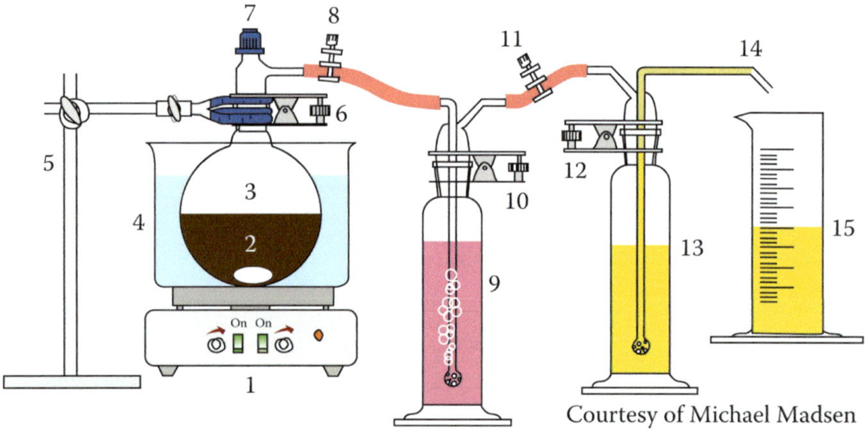

FIGURE 2.8
Scheme of a volume displacement measurement system built with standard laboratory glassware, suggested by Dr Michael Madsen. (1) Magnetic stirrer with thermostatic plate, (2) Bottle with magnetic bar for stirring, employed as digester. (The picture shows a laboratory balloon, but in our opinion an Erlenmeyer flask with flat bottom would be more convenient), (3) Digester's headspace, full with biogas, (4) Water bath for maintaining a uniform, constant temperature, (5) Support with clamps, (6) Collar for the flask's adapter, (7) Flask adapter with valve for sampling biogas, (8) Rubber tube with flange or hose clamp, (9) Drechsel bottle (could also be a Muencke bottle or any other similar device for washing gas), filled with NaOH solution and pH indicator (typically phenolphthalein, becoming colorless at pH < 8), (10) Collar for the Drechsel bottle, (11) Rubber tube with hose clamp, (12) Collar, (13) Drechsel bottle filled with colored water to facilitate the readings (the author of Figure 2.8, Dr. M. Madsen, recommends methyl orange but any other colorant will be suitable, since the scope of coloring the water is just for making it easier to read the graduated cylinder.), and (14) Graduated cylinder. (Picture with Wiki Commons license.)

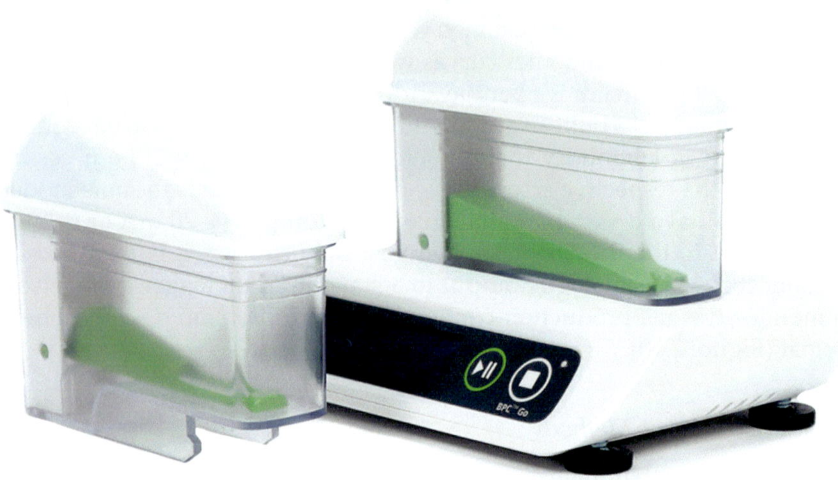

FIGURE 2.9
BPC Go flow meter and datalogger. (Photo courtesy of BPC Instruments AB.)

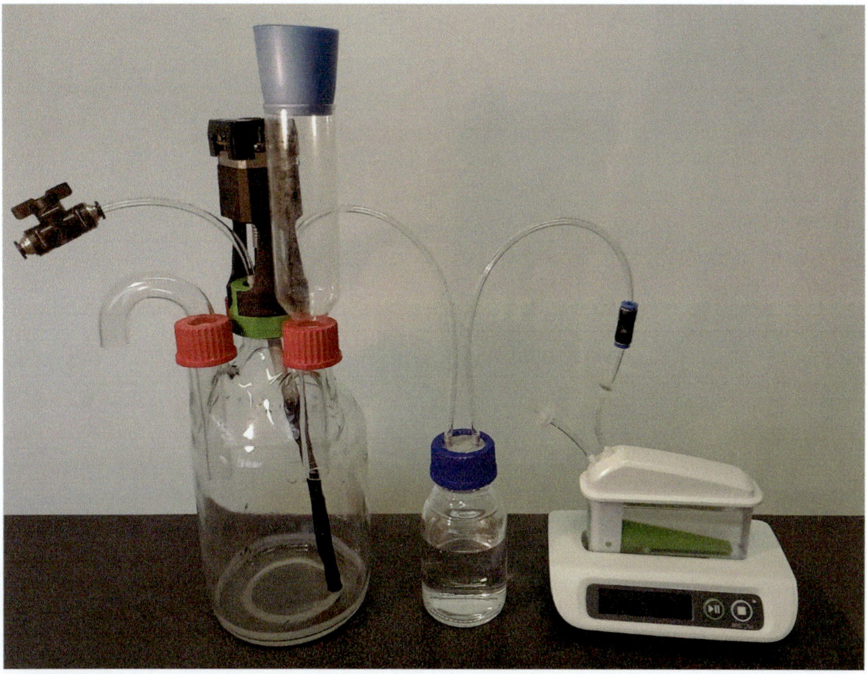

FIGURE 2.10
Low-cost measurement system built by the author with a semi-batch glass reactor, a bottle with NaOH solution, and a BPC Go flow meter and datalogger.

The cap has a small hole that connects the gas in the bottle to a manometer via a tube passing through. Figure 2.11 shows the concept.

Since the bottle's headspace volume is constant, because water is incompressible, the ratio between the gas pressure and the volume of biogas produced by the AD, normalized at 0°C and 101.3 kPa, is calculated with the state equation of perfect gases. Under such measurement conditions, this is expressed as:

$$P_r \cdot V_r = m \cdot R \cdot T$$

where

P_r = pressure inside the reactor [Pa]
V_r = volume of gas contained in the reactor [m^3] = $V_{bottle} - V_{sludge}$
m = mass of gas produced by the anaerobic fermentation [mol]
R = universal constant of gases = 8.314 m^3 Pa/mol·K
T_r = Absolute temperature of the gas within the reactor [K] (it is assumed to be the same temperature of the incubator)

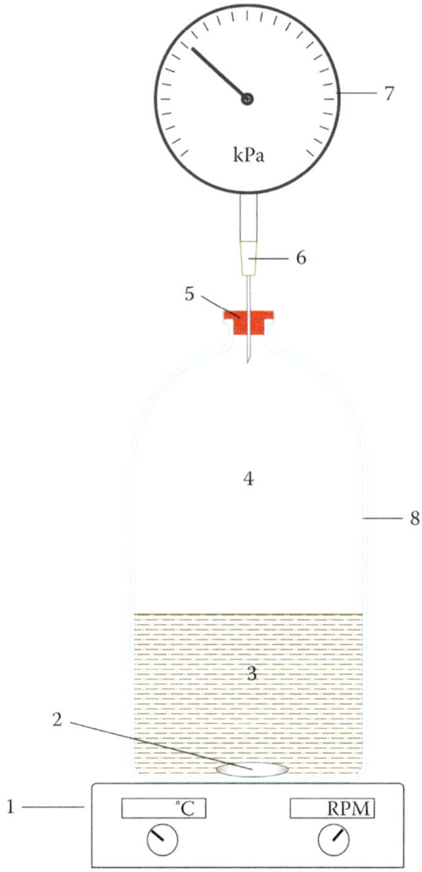

FIGURE 2.11
Elementary barometric system. (1) Heated magnetic stirrer (optional; the bottle can be placed in an incubator and stirred by hand a few times a day), (2) Magnetic bar for stirring, (3) Slurry under fermentation, (4) Collection space of the gas, (5) Hermetic cap, clamped to the bottle's neck, (6) Tube piercing the cap, (7) Manometer, and (8) Glass serum bottle, with mouth suitable for clamping the cap. (Drawing by the author.)

Since the measured volume of gas must be referred to normal conditions (273.15 K and 101.3 kPa), the former equation becomes:

$$101{,}300[\mathrm{Pa}] \cdot V_\mathrm{N}\left[\mathrm{Nm}^3\right] = m[\mathrm{mol}] R \cdot 273.15[\mathrm{K}]$$

Our variable is the value V_N.

Under the test conditions, the gas contained in the reactor must satisfy the following thermodynamic state equation:

$$P_\mathrm{r}[\mathrm{Pa}] \cdot V_\mathrm{r}\left[\mathrm{m}^3\right] = m[\mathrm{mol}] \cdot R \cdot T_\mathrm{r}[\mathrm{K}]$$

The quotient between both members of the equations corresponding to the test and the normalized conditions allows simplifying the product $m{\cdot}R$. Hence:

$$\frac{P_r \cdot V_r}{101{,}300 \cdot V_n} = \frac{m \cdot R \cdot T_r}{m \cdot R \cdot 273.15}$$

So, the volume of gas produced by the AD, normalized to 0°C and 101.3 kPa, can be calculated with the measured values P_r and T_r and with the known headspace volume of gas in the reactor, V_r, with the following formula:

$$\frac{P_r \cdot V_r \cdot 273.15}{101{,}300 \cdot T_r} = V_n$$

The formula indicates that, since V_n is derived from a calculation based on three measured quantities, the relative error of the calculated V_n will be equal to the sum of the individual relative errors of each measured variable, i.e., P_r, V_r, and T_r.

Furthermore, the pressure in the headspace is the sum of the partial pressures of all gases contained in it (including the N_2 or other gas employed to flush the air, methane, carbon dioxide, and trace gases) and water vapor. The latter is proportional to the temperature, so the water vapor at the test temperature must be discounted before applying the formula for perfect gases (dry gases) already described. If employing a blank to obtain the net methane production by difference, then the systematic error induced by the moisture is null, on condition that the headspace volume is the same both in the blank and in the sample reactors. The following reasoning shows why:

$$G = S - B$$

where

G = net gas production of the substrate (expressed as pressure increment)

S = gross gas production of the sample reactor

B = gross gas production of the blank reactor

Observe that, according to Dalton's Law of perfect gases, both S and B are the sum of the partial pressures of gas in the headspace volume, S_g and B_g, and water vapor, S_v and B_v; hence:

$$G = S_g + S_v - \left(B_g + B_v\right)$$

If the headspace is the same in both reactors, then $S_v = B_v$, so the systematic errors cancel each other and G is actually the net biogas production. The reasoning above does not apply to barometric instruments that vent part of the gas when the internal pressure exceeds a safety limit. In this case, water vapor will be released together with the gas; however, the number of gas ventings differ between the sample and the blank, so the volumes of vapor will not cancel out when calculating the BMP.

Volumetric counters also release gas at each cell opening, similar to the automatic vented barometric systems described above. Therefore, it is necessary to correct the moisture in the gas for both the sample and blank reactors.

Practical Conclusion

Having the same headspace in both the sample and blank reactors is a necessary condition for minimizing the error of biological tests. The sensitivity to the error induced by the moisture content in the gas differs between barometric and volumetric methods: in the former, the error induced by eventual differences in the headspace is significant, as the headspace is large. In volumetric systems, the headspace is small, and the partial volume of water vapor (at 38°C) is less than 6% of it, resulting in a systematic error that is below the detection limit of most volumetric measurement devices. Some measurement systems, such as the third-generation products of BPC Instruments AB (AMPTS III, BRS III, GE III), have a built-in function to account for water vapor and provide the normalized volume of dry gas.

As was the case for volumetric methods, apparatuses based on the barometric method can be either self-constructed (which means procuring adequate sensors, assembling the reactors, and calibrating each of the latter) or purchased on the market as pre-assembled kits. In general, the latter are unstirred devices, in which the pressure variations are measured with electronic sensors and processed by software. Figure 2.12 illustrates an example of self-built reactors equipped with mechanical manometers, and Figure 2.13 shows an example of a barometric system, typically used for biological oxygen demand (BOD) measurements and occasionally for biochemical methane potential (BMP) analysis, produced by VELP Scientifica srl.

The disadvantages of barometric methods compared to volumetric ones can be summarized as follows:

1 Because of their nature, barometric methods only measure the total biogas produced. Hence, it is necessary to perform an additional measurement to determine the methane concentration. This adds an error (2% to 5%) to the final value; consequently, they are less accurate compared to volumetric methods that measure the net methane production directly through a CO_2 trap.

2 Barometric systems seldom have stirrers (in the best case, they have just a magnetic bar stirrer). Stirring is a crucial factor in AD; therefore, this fact leads to discrepancies between the production measured in the laboratory and the actual yield of the plant (see Chapter 6). Please note that biogas plants work at (almost) constant pressure, i.e., in conditions that are more similar to those of volumetric methods.

FIGURE 2.12
Example of homemade barometric reactors employed by the company Biobooster srl (Italy) during their start-up research. (Photo by the author.)

3 CO_2 and CH_4 have different solubilities in water. As the pressure in the reactor increases, a fraction of the produced CO_2 dissolves in the aqueous phase. This distorts the measured values in two ways: On one side, the proportion of carbonates in the slurry changes, influencing the alkalinity and, ultimately, the kinetics of the organic matter's degradation. On the other hand, as CO_2 dissolves in the slurry, the concentration of CH_4 in the collected gas increases, giving the false impression that the biogas quality obtained is higher than it actually is.

4 Barometric methods are more popular in the sector of *aerobic tests*, for instance, the determination of the BOD of sewage water. They were indeed born for this scope and later adapted to measure BMP.

5 Sometimes the samples produce more biogas than foreseen; hence the reactor's pressure rises too much and, if for any reason the safety valve does not release the excess gas, we risk finding our laboratory as shown in Figure 2.14.

6 Some barometric systems automatically release overpressure, and the total biogas is then computed as the sum of n independent measures of pressure and gas composition. Since each measure has its error

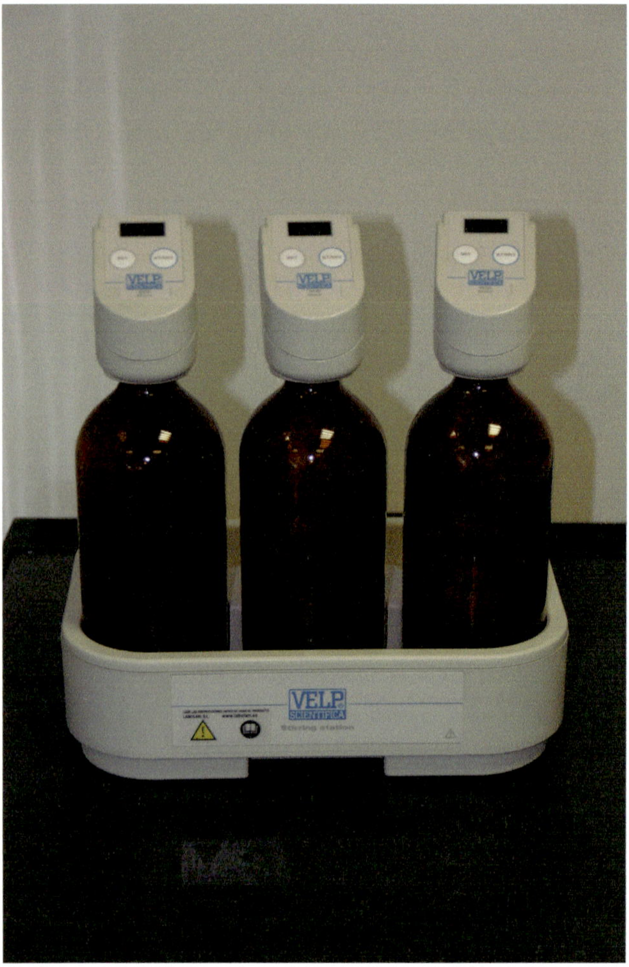

FIGURE 2.13
Barometric bottles, a device originally conceived for *aerobic* respirometric measures. Such systems are typically placed in a thermostatic cabinet and lack stirring. The communication between the pressure sensor and the datalogging unit is wireless, which introduces an additional weak point to the system: Li-ion batteries have a shorter lifespan when operating at high temperatures, leading to frequent interruptions and data loss during tests. (Photo by the Author, with thanks to Dr. Diana Castro Anca, SOLOGAS, Spain.)

margin, the total final absolute error is *n* times the individual absolute error.

7. Barometric systems sending the pressure signal to the datalogger via wireless interface sometimes lose the connection because the battery in the sensor unit gets exhausted before the end of the test. This is particularly common in systems where bottles with wireless sensors are

FIGURE 2.14
Unpleasant consequences of a reactor's explosion caused by overpressure. (Photo by Åssa Jarvis, 2010.)

placed in thermostatic cabinets at 38–45°C, as the battery's life drops drastically with rising temperatures, especially in the case of Li-ion batteries. Losing data during a test makes the results unreliable.

2.2.2 Automatic Measurement Systems: AMPTS and BRS

The techniques for measuring the volume of biogas produced by AD have remained almost unchanged since the first experiments conducted by Alessandro Volta with the "air that burn", and Pasteur's respirometers for studying the fermentation of glucose. Over the last two decades, the growth of the biogas industry has increased the need for maximum standardization of tests and simplification of procedures, aiming to minimize manual work and human errors. The most modern systems currently available on the market (2025), unlike their predecessors, are specifically designed to measure anaerobic fermentative processes and hence employ digital technology, containing a series of innovations that make them particularly useful for the biogas

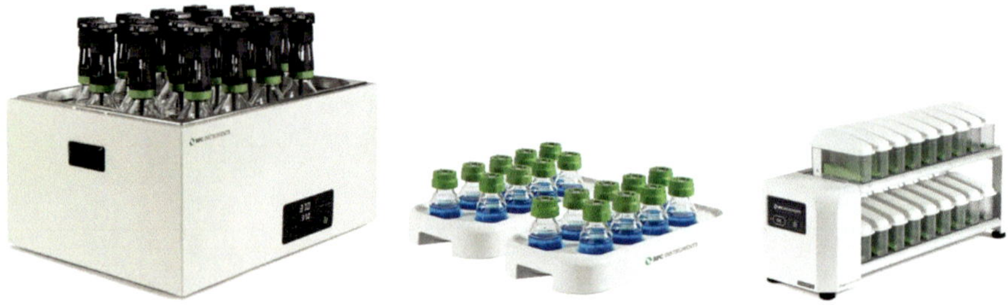

FIGURE 2.15
The AMPTS III. (Photo courtesy of BPC Instruments AB.)

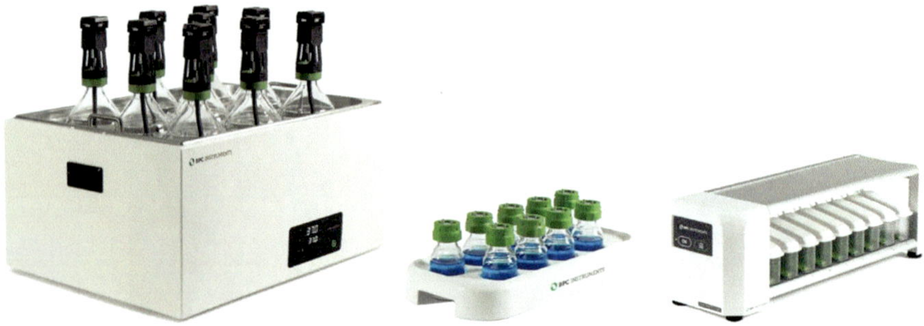

FIGURE 2.16
The AMPTS III Light, a smaller and cheaper version of the AMPTS III, which is more suitable for industrial use. (Photo courtesy of BPC Instruments AB.)

plant manager. The current market standards are the AMPTS III (Automatic Methane Potential Test System of third generation) and its smaller version, called AMPTS III Light. Both are specifically designed for performing batch tests, whereas the BRS III (third-generation Bio Reactor Simulator) is a lab-scale system for continuous process simulation. Both the reactors and the measurement methods of the said instruments have been standardized to the maximum extent, yielding highly repeatable results without the need for skilled personnel in the laboratory (Figures 2.15, 2.16, and 2.17).

The said instruments share different functional features and even their external appearance, but their scopes in analyzing fermentative processes are pretty different. Their shared features are:

1 Measurement method: liquid displacement counter, with interchangeable measurement cells (2 mL and 9 mL).
2 Real-time normalization of the measured volume: all of them include pressure and temperature sensors in their electronic circuits and a normalization software that calculates the relative moisture of the gas and

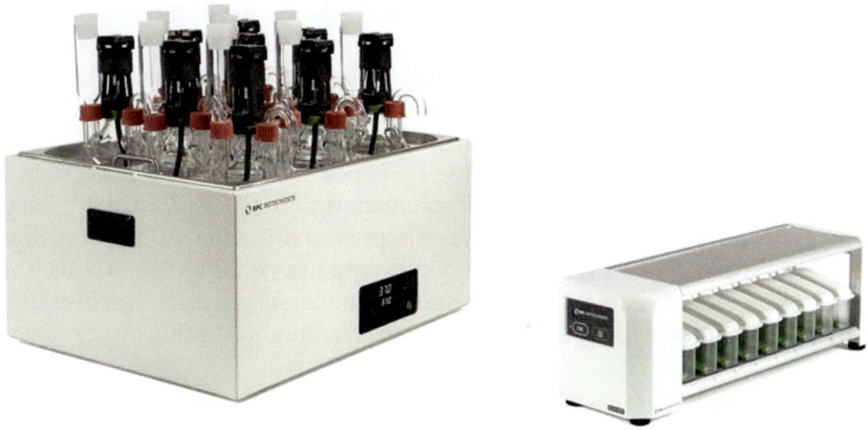

FIGURE 2.17
The BRS III. When performing continuous process simulation, it is customary to measure the total biogas production. Consequently, the BRS III has no CO_2 absorption flasks. (Photo courtesy of BPC Instruments AB.)

compensates for the composition of the flush gas in the reactor's headspace to correct the reading.

3 Highly automated calibration protocol (maximum error ±1% of the reading).
4 No need for periodical recalibrations.
5 Remote access to the data through LAN or Internet—no need to go physically to the laboratory to check your experiment.
6 Interactive software based on the Hyper Text Markup Language (HTML) format. This means that the interface for accessing the instrument can be any Internet browser. Such a feature has the following advantages: no need to have a dedicated PC, since it can be accessed even through a smartphone, nor is it necessary to install any special software, since its operation is independent of the operating system.
7 Data output both graphically (on the PC's screen) and in numerical format (Excel [XLSX] and comma-separated values [CSV] tables, easily readable with any spreadsheet).

From the point of view of the biogas plant's manager, the most crucial aspect to know about these instruments is their practical application. The AMPTS III has 18 channels measuring the gas flow from the corresponding 1,000 mL batch reactors. It is an instrument designed to measure the biomass's BMP (for instance, to determine its market price based on its effective energy potential). Therefore, the AMPTS III is built and calibrated to directly measure the net amount of methane produced by cleaning biogas through filters containing caustic soda, which eliminates carbon dioxide. As shown at the beginning of this chapter, measuring a given magnitude *directly* provides higher accuracy of the results than measuring several variables and then calculating the result using a formula. The differences between AMPTS III and its Light version are the number of channels (9 instead of 18) and the reactors' size (2,000 mL), which facilitates the preparation of the feedstock sample for use in the industrial environment. The BRS III is an instrument designed for performing long-term continuous tests, and its software is prepared to store and process the relevant parameters within this scope: the organic loading rate (OLR), the hydraulic retention time (HRT), and the instantaneous gas flow. The BRS III is particularly suitable for analyzing dynamic operation conditions:

1 To simulate transient states (e.g., the possible strategies for starting a new digester or the restart after a biological collapse).
2 To analyze possible modifications to the operation of the plant (e.g., to check if, after changing the digester's diet, it may be convenient or not to switch from mesophilic to thermophilic operation, and how to carry out the said change without altering the productivity of the plant).
3 To simulate the operation of a multistage plant, or the influence of a variable OLR, or the eventual toxicity of a new feedstock.

Summarizing: The AMPTS III Light is suitable for managing one or two biogas plants and for determining the fair price of biomass. The AMPTS III is ideal for research purposes, managing biogas plants that treat a larger variety of feedstock, certifying the quality and methane yield of waste organic matter and commercial feedstock, and for centralized management of several plants. In the last case, it is possible to manage easily from two to five plants, assuming these are within 100 km of the laboratory. Bigger distances make the transportation of samples more expensive and complicated, especially if these are perishable biomass. The operation of BRS III is outside the scope of this book, being mainly conceived for scientific and industrial research (universities, constructors of biogas plants, specialized consultancy, engineering companies, and owners of very complex biogas plants working with variable loads). Continuous digestion tests are explained in more detail in the second book of the same author in this collection: *Optimising Biogas Plants, Squeezing Profit out of your Digester*.

2.2.3 Reactors for Biological Tests: Which Are Better?—Big or Small Ones?

The debate over the reactor's dimensions is always a topic of discussion. Some manufacturers and academics will assert that "the bigger, the better". Some individuals will affirm that "20 and 50 L is the standard in Germany" (a false argument, because the German norm VDI 4630 provides a guideline to choose the size of the reactor, depending on the test's scope, but does not state a concrete volume; see Chapter 6). We will see in Chapter 3 that, when employing volumetric methods, the metrological error in the determination of a feedstock's BMP is independent of the reactor's volume. In Section 2.2.4, we will introduce the errors generated by the test conditions, which are independent of the reactor's volume (once again, only if volumetric methods are adopted). The belief that the test's accuracy is directly related to the reactor's volume stems from the old technologies for measuring gas flows with instruments similar to those used in domestic natural gas counters, characterized by fair accuracy within a given flow range, but lacking sensitivity in the case of extremely low flows. Modern laboratory flow meters like AMPTS and BRS have a high sensitivity, in the range of 2–9 Ncm^3, depending on the flow cell employed. Hence, it is unnecessary to employ reactors larger than 2 L (batch) or 5 L (continuous), as the gas flow from the most frequent substrates will be high enough to avoid detection problems. Big volume reactors are required for some particular cases, like the research on the AD of very refractory substrates (for instance, pesticides and biodegradable plastics) and low concentration wastewater, or in the case of extremely heterogeneous and coarse substrates, like organic urban waste. In the first case, a large reactor is necessary to produce a quantity of gas large enough to be measured in a reasonably short time, unless an instrument with high sensitivity is employed to measure the gas flow. In the second case, a big reactor turns out to be more suitable for digesting a volume of sample large enough to be considered "representative" of a full truck of very heterogeneous garbage, or because the particle size requires a feeding tube of large diameter and hence a proportionally bigger reactor.

In the special case of managing an agricultural or industrial biogas plant, reactors from 500 mL to 2 L are more than enough to cover all possible routine tests. At this scale, the reactor's size influences the precision of the test rather than its accuracy. Using 1 or 2 L reactors is the best choice because the size of the samples will be bigger, i.e., the sample will be more representative, leading to smaller dispersion between replicates (i.e., a smaller coefficient of variation, CV, defined as the standard deviation among replicates divided by the average gas production). The Italian norm on BMP assay, UNI/TS 11703:2018, defines that the minimum quantity of inoculum for a reliable test must be 400 mL, and the consequent reactor's size will then depend on the method adopted, volumetric or barometric.

Figure 2.18 shows an old bank of reactors, 15 L each, belonging to CIEMAT, Madrid, nowadays employed only for some special research on scarcely degradable substrates.

2.2.4 The Most Frequent Errors in the Measurement of Small Gas Flows

2.2.4.1 Normalization

If an instrument with built-in normalization in real time is not available, measuring correctly requires reading the volume of gas (or the internal

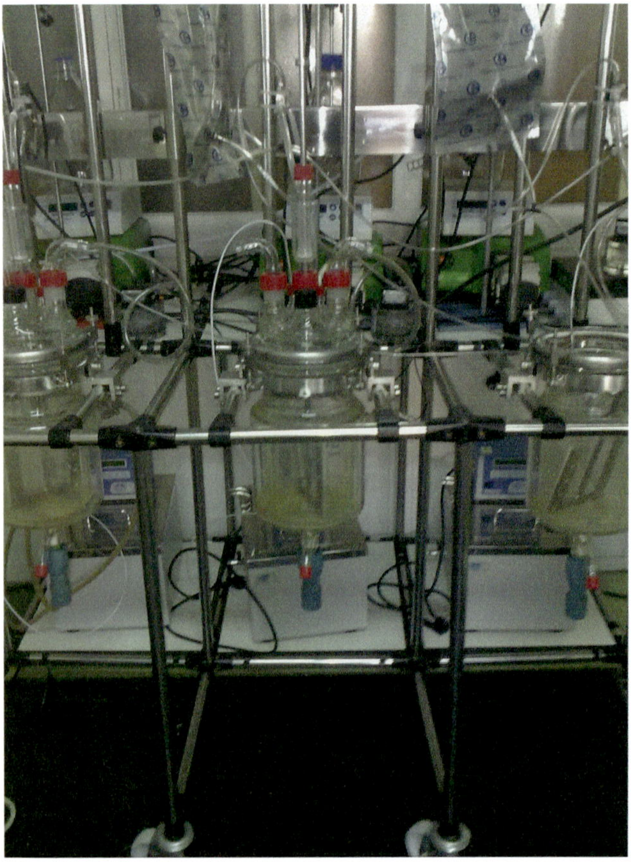

FIGURE 2.18
15-l reactors bank in use at CIEMAT, Madrid. (Photo by the author.)

pressure of the reactor, if a barometric method is employed) and also the ambient pressure and temperature in that moment, allowing to apply the state equation of perfect gases to obtain the normalized value accordingly. Adopting this method implies that the relative errors of the gas volume (or pressure) measurement system, of the ambient temperature (or the temperature of the incubation chamber of the reactor), and the atmospheric pressure will add up. It is a frequent error (especially in older publications) not to perform these corrections when taking each reading, and assuming a standard pressure of 101.3 kPa and a temperature of 20°C as constant laboratory conditions for all readings. This introduces an additional error of 6%–12%, as atmospheric pressure can vary by more than 60 mbar in a month, and temperature, even in a research laboratory, can fluctuate by more than 6°C, especially on weekends (Strömberg et al., 2012).

As explained in Section 2.2.1.2, the formula for the normalization of volume in the case of measures taken with barometric methods is:

$$\frac{P_r \cdot V_r \cdot 273.15}{101.300 \cdot T_r} = V_N \text{ correct}$$

With the same reasoning, it is possible to deduce that, in the case of volumetric methods, the normalization formula becomes:

$$\frac{P_{atm} \cdot V_{mis} \cdot 273.15}{101{,}300 \cdot T_r} = V_N \text{ correct}$$

where

P_{atm} = atmospheric pressure (measured with an ambient barometer) at the moment of reading the volume of gas (Pa)

V_{mis} = measured volume of gas (m^3)

T_r = room temperature in K (273.15 + T in °C)

On applying the rules of error propagation, even in the case that each single reading is normalized on the room conditions of the moment, the normalization of the gas volume introduces a relative error equal to:

$$e_{V_N} = e_{\text{barometer}} + e_{\text{thermometer}} + e_{\text{gas counter}}$$

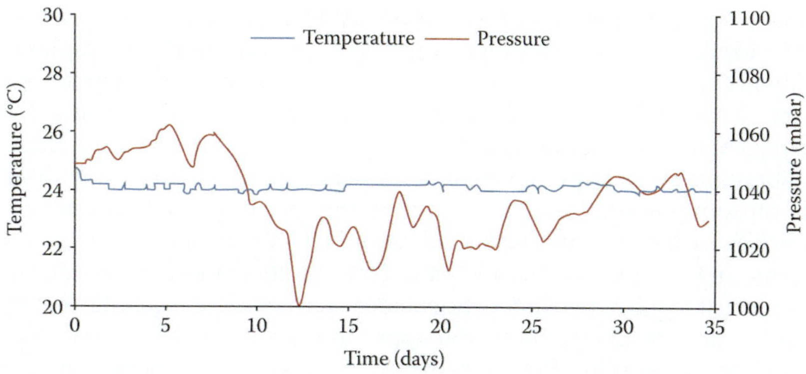

FIGURE 2.19
Variations of P_{atm} and T_r during a typical month in the laboratory of the University of Lund. (Courtesy of Prof. Jing Liu.)

If the experimental set does not have a CO_2 trap, then an additional measure is necessary—the composition of the biogas. Hence, the total relative error becomes:

$$e_{V_N} = e_{\text{barometer}} + e_{\text{thermometer}} + e_{\text{gas counter}} + e_{\text{gas analyzer}}$$

2.2.4.2 Correction of the Moisture

The headspace of a digester is filled with saturated biogas, i.e., having a relative moisture content of 100%. Under the usual temperature conditions of biogas plants (in the range of 36°C–50°C), the total volume of gas contains 6%–10% of water vapor. In general, the instruments and simplified formulas presented in the preceding paragraphs do not account for the effect of moisture, which can induce further error. Remember that the state equation of perfect gases described in the former paragraphs, although being quite accurate in the range of pressures and temperatures near ambient, is only valid for *dry gases*. As explained in Section 2.2.1.2, moisture induces a systematic error that is automatically cancelled in barometric methods, provided that the headspace in both the blank and the sample reactors is the same and no gas is vented during the process. In all remaining cases, it is necessary to adjust the reading for moisture.

2.2.4.3 Elimination of Carbon Dioxide

When employing instruments without caustic soda filters (typically barometric instruments, but also some volumetric ones), the measured value represents the *gross* amount of biogas produced, having extremely variable methane and carbon dioxide proportions (the assumption of 60% CH_4 and 40% CO_2 is purely conventional). To obtain an accurate value of the methane concentration, it is then necessary to analyze the gas using a gas chromatographer or another type of gas analyzer. Gas chromatographers are highly accurate but too expensive for a laboratory in a biogas plant – and for many research facilities too! A cheaper alternative is an instrument with electrochemical cells (affordable cost, easy operation, but requiring frequent recalibrations and a large volume of sample gas), or an instrument with solid-state sensors (intermediate cost, but requiring periodical substitution of the sensors every 2 or 3 years and recalibration, and about 1 liter of gas sample). Determining *a posteriori*, the methane concentration induces an additional error, since it requires two independent measurements—the quantity of biogas produced and its composition. The application of a coefficient "from the literature" (e.g., assuming 55% of methane if the digested biomass is composed of polysaccharides, and 60% if it is composed of proteins) induces an error too coarse to be acceptable. It is more convenient to plan the experiments in such a way that a caustic soda filter ensures measuring the *net amount* of methane produced by the reactor. It is furthermore advisable to discard the caustic soda solution at the end of each assay and to begin the next with a fresh solution, thereby avoiding the crystallization of sodium carbonate (Na_2CO_3) in the filter. Another inexpensive method for determining the composition of biogas is the "syringe method", described in Chapter 5, Section 5.2. The syringe's accuracy limits the accuracy of the gas composition measurement.

Finally, using the AMPTS III Duo provides a continuous datalogging of the approximate biogas composition. AMPTS III Duo is an instrument that comes with two gas measurement units (called BPC Core), providing two readings: the total biogas and the net methane volume. This special setup and the analysis of its error propagation are described in detail in the book already cited (Optimising Biogas Plants, CRC Press, 2025).

2.2.5 Auxiliary Physicochemical Tests

2.2.5.1 Measurement of the pH and the ORP

Two methods exist for measuring pH: by using paper strips impregnated with substances that change color depending on the pH level, or by using a pH meter. The first method is not advisable in biogas plants, because both sludge and digestate tend to color the paper brown or black, making the reading impossible. For *in situ* measurements, it is advisable to employ a digital pH meter, even a basic pocket model like the one shown in Figure 2.20, which measures pH, temperature, and electrical conductivity.

We have seen in Chapter 1 that the pH value is only a coarse control parameter, because it does not provide much information about the actual state of the AD process. It is anyway interesting to monitor it; hence, the pH meter must be kept perfectly efficient. It is then mandatory to purchase the cleaning and calibration solutions for the electrodes, calibrate the instrument periodically, and follow the manufacturer's instructions. The instrument must be thoroughly washed after each use, so as to avoid the formation of dirt deposits on the electrode.

All those prescriptions already said about pH meters apply to ORP meters too.

2.2.5.2 Determination of the *DM and VS* of Biomass

The determination of the DM of any substance is relatively easy to carry out. All the instruments required are a scale—if possible, with a 1 mg error margin or less—and an electric oven (even a small "convection oven" is sufficient, as the temperature must not exceed 110°C). A halogen moisture analyzer is a special type of laboratory scale consisting of a system for drying the sample with a halogen lamp and the necessary electronic circuitry to automatically sense the weight variation and determine the end of the test (Figure 2.21). Such an instrument is easier to use and saves a significant amount of time,

FIGURE 2.20
Example of pocket combo meter (pH, temperature, electric conductivity, total dissolved solids). (Photo by the author.)

as it performs the test automatically. Using an oven will usually take longer than using a halogen moisture analyzer. If you opt for the oven method, the easiest way is to place the samples in the oven during the evening and leave them overnight. In the second case, the test can last between 15 and 60 min, depending on the moisture content and grain size of the sample.

When neither an oven nor a scale is available, it is convenient to purchase a moisture analyzer with an acceptable error margin (±1 mg or less) so that it can also be used for ash determination. A moisture analyzer occupies less space and consumes less energy than an oven, allows for quicker and automated tests (less probability of human errors), and can also be used as a laboratory scale. Many models are available on the market, each with different functions, accuracies, sensitivities, and, of course, prices. As a rule, avoid instruments with an error margin (sensitivity) higher than 1 mg.

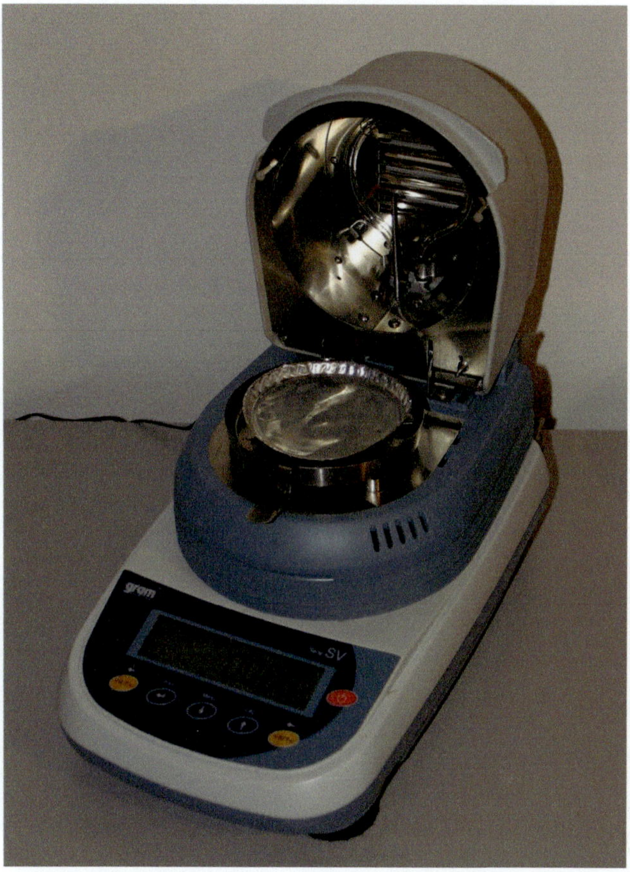

FIGURE 2.21
An example of a halogen moisture analyzer for the automatic determination of DM and moisture, and for laboratory precision weighing in general. (Photo by the author.)

The determination of the VS requires three weightings: the wet sample, the dry sample, and the ash that remains after the calcination. A muffle oven is needed to calcine the biomass samples, i.e., to reduce them to ashes. The standard procedure is to heat the muffle to 550°C, i.e., the self-ignition temperature of any organic matter. The sample of dry biomass can be introduced in the oven once its temperature is stabilized. All carbon, hydrogen, oxygen, and nitrogen compounds will volatilize, while the ash will remain in the plate or crucible. The direct combustion of the sample on a flame is not acceptable for several reasons, mainly because a flame's temperature is almost double the self-ignition temperature of biomass, and will partly volatilize the ash, giving a wrong result. Figure 2.22 shows a model of an electric muffle oven. When considering the purchase of a muffle oven, it is essential to remember that two different technologies exist: those with a heavy refractory inner

FIGURE 2.22
Example of a muffle oven. The small ceramic tube on the top allows the evacuation of fumes. Air enters through some holes placed under the door's insulation. (Photo by the author.)

lining and those built with a light alumina silicate, featuring only the floor of the inner chamber in hard ceramic material. In the first case, the muffle will have higher thermal inertia (hence higher temperature stability) and its inside will be more resistant to abrasion, small shocks, etc., but will consume more energy to reach the required temperature. The disadvantage of this technology is that its startup involves a series of heating cycles with increasing final temperatures, followed by slow cooling. This time-consuming procedure is mandatory to dehydrate the refractory lining and mortar; otherwise, the vapor pressure could cause the ceramics to burst and get damaged. The ovens built with an inner lining of alumina silicate are lighter and do not require a start-up procedure. However, the material is fragile and soft, so it must be treated delicately, avoiding scratches and shocks. Since the second type of oven is produced from a single insulating block characterized by low thermal mass, it heats up quickly and, in general, consumes less energy than refractory ovens. In any case, before purchasing any given model, it is essential to verify that it has a fume escape duct or hole, as this feature is not always standard.

2.2.5.3 Determination of the COD of Sludge and Liquid Substrates

Sludge (in general, the content of a digester, the digestates, liquid manure, industrial slurry, and similar substances) usually contains less than 10% solids. Fiberless slurry (for instance, sludge from wastewater treatment plants, whey, blood...) usually contains too little solids for the easy determination of the organic matter with the method of the VS, because of the large quantity of water to evaporate or because volatile organic compounds (VOC) are dissolved in the water. In such cases, it is more convenient to measure the chemical oxygen demand (COD). This is a highly standardized chemical test, and pre-dosed kits exist in the market for using them in automatic instruments, allowing even unskilled operators to perform accurate and reliable tests. The most widely adopted method is known as photometry. Its working principle is based on the experimental fact that the absorption of light is a phenomenon of interaction between electromagnetic waves and matter. When a light beam travels across a substance, part of the electromagnetic radiation can be absorbed by atoms, molecules, or crystalline lattices.

Suppose pure absorption takes place (neither reflection nor refraction). In that case, the absorbed fraction of light depends both on the length of the optical path across matter and on the physicochemical properties of the substance, according to the Lambert–Beer law:

$$A = -\log\left(\frac{I}{I_o}\right) = \varepsilon_\lambda \cdot c \cdot d$$

where

A = absorbance (a dimensionless coefficient depending on the substance)

I = Intensity of light at the exit of the optical path

I_o = Intensity of light at the entry to the optical path

ε_λ = molar extinction coefficient at the wavelength λ

c = molar concentration of the substance

d = optical path across the substance

The working principle of a photometer is sketched in Figure 2.23.

The liquid under test is dosed into the vial, where it reacts chemically with the content, changing the color of the solution. After a thermal reaction at high temperature, the vial is introduced into the spectrophotometer, where it is positioned between a light source of a given wavelength and intensity and

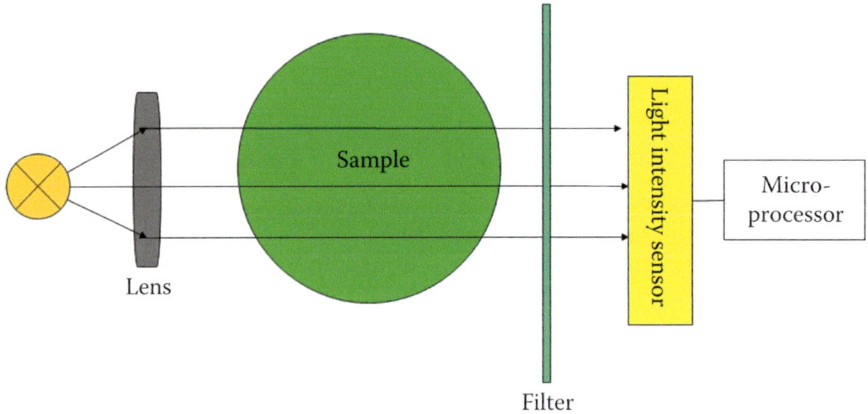

FIGURE 2.23
Sketch of the photometer's working principle.

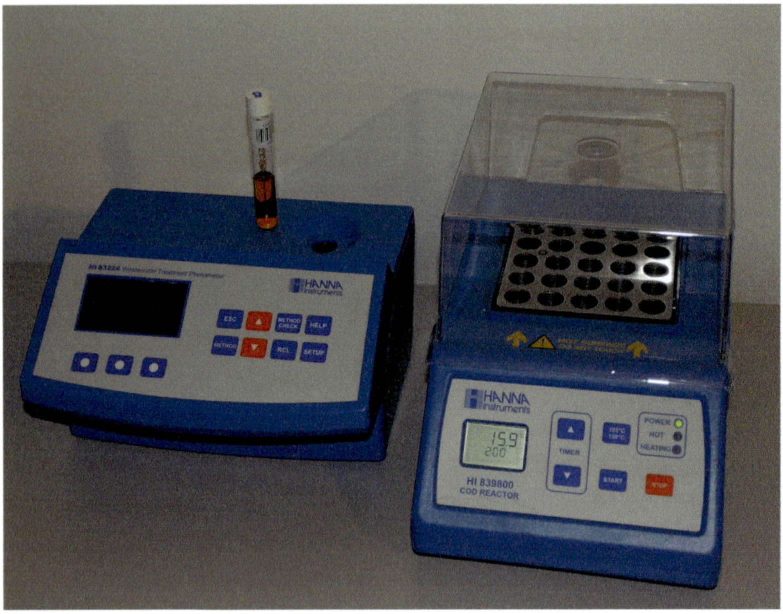

FIGURE 2.24
Multiparameter photometer for wastewater, cuvette with barcode for automatic measure range setting, and electric furnace for pre-treating the cuvettes—a pre-treatment required by some tests like COD and Ammonia Nitrogen. (Photo by the author.)

a light sensor. The difference between the light intensity across the vial and a reference vial called the blank (loaded with distilled water) is then digitally processed by a microprocessor, and the reading appears in the instrument's display.

Figure 2.24 illustrates a model of a photometer equipped with vials featuring automatic barcode identification. In this way the operator needs only to introduce in the vial the prescribed amount of sample (using for said operation an automatic pipette that samples exactly the necessary amount), then shake it gently, place it in the electric heater for the prescribed time and, after cooling, take the reading in the instrument. The photometer will then read the vial's barcode, ask for confirmation of the method, and proceed to measure the COD (or the parameter identified by the barcode). This highly automated procedure eliminates the possibility of human errors associated with traditional photometers, such as taking a vial for the measurement of total nitrogen and reading the measured value as COD.

Photometers can be either multiparameter or monoparameter. The most interesting parameters for the biogas plant manager are COD, total nitrogen, ammonia nitrogen, and, if possible, also total phosphorus. The possibility of also measuring total nitrates and total potassium can be interesting for biogas plants that employ the digestate as fertilizer.

2.2.6 The *In Situ* Laboratory for the Real-Time Control of the Biological Process

From a practical point of view, the minimum laboratory equipment for monitoring the biological processes in a biogas plant is listed below:

1 Automatic moisture analyzer scale (±1 mg resolution) or analytic laboratory scale (±1 mg or ±0.1 mg resolution), maximum weighing range 100 g.
2 Muffle oven (at least 2.5 L capacity, temperature control at least up to 650°C, accuracy ± 10°C).
3 Combo pH meter, pocket type (pH/temperature/electric conductivity).
4 Oxidation reduction potential (ORP) meter (optional).
5 Accessories: plates, crucibles, tongs, high-temperature protection gloves, scissors, brushes for cleaning bottles and flasks, several funnels of different sizes, a stainless-steel sieve with a 5-mm to 10-mm mesh, safety eyeglasses, single-use gloves, and a safety eye wash kit.
6 At least three kits for measuring the volume of gas produced by the AD (at least one blank and two replicates), including reactors with a stirrer, a thermostatic bath, a caustic soda filter, and a gas flow measurement device (preferably with a built-in gas volume normalization function).
7 Graduated cylinders (preferably class A) of 50, 100, and 250 mL.
8 Photometer (at least for COD), automatic sampling pipette of the volume recommended by the instrument's manufacturer (usually 200 µl), kit of pre-dosed vials, table heater for the conditioning of the vials

(this one is optional; the muffle oven could be used instead, but its temperature control is less accurate and consumes more energy).

9 Working plane: at least two tables lined with Formica or melamine sheet, 60 cm ×120 cm. The surface must be rigid and stable.

10 A washbasin (stainless steel or porcelain), preferably equipped with a dripper, connected to a water line and drain sink.

11 A technical scale having maximum weighing range 4 kg and 0.1 g error margin.

12 Consumables: single-use gloves, soap and detergents, caustic soda (NaOH), distilled or demineralized water, wide-mouth plastic bottles for slurry samples, reference substrates for biological tests (acetic acid or anhydrous sodium acetate, starch, gelatine, cellulose, optionally also propionic and butyric acids).

A small room (12–15 m^2), a box inside an industrial building constructed with plasterboard or any other type of drywall panel, a prefabricated box, or a 20′ container with insulated walls (such as those used as temporary offices in construction yards) is sufficient for installing the laboratory. It is essential that the door has a key or lock to prevent unauthorized access, and that the working space is well ventilated, either through windows or an electric aspirator (especially during the calcination of biomass samples with the muffle

FIGURE 2.25
Model laboratory for BMP analysis in the biogas plant. (1) Washbasin with dripper, (2) Muffle oven, (3) AMPTS II, (4) Photometer and electric heater for the vials, and (5) Moisture analyzer scale.

oven). The total electric power available must be at least 3 kW (typically, the muffle oven requires approximately 2 kW). Figure 2.25 shows a model laboratory equipped according to the former indications, installed in a room measuring approximately 12 m^2.

Bibliography

Arrojo, D., and Dampé, I., *Medidas eléctricas*, notes of the course on electric measures, Universidad Nacional de La Plata, Buenos Aires, Argentina, 2013.

BIPM (Bureau International des Poids et Mesures), Evaluation of measurement data. *Guide to the Expression of Uncertainty in Measurement, JCGM 100*, First edition, 2008.

Bolzonella, D., *Monitoring and Lab Analysis—Assays*. Lecture of Jyvaskyla Summer School, Sweden, 2013.

Esteves, S., Devlin, D., Dinsdale, R., and Guwy, A., Performance of various methodologies for assessing batch anaerobic biodegradability, International IWA Symposium on Anaerobic Digestion of Solid Wastes and Energy Crops, Vienna, Austria, 2011.

Ficara, E., Scaglione, D., and Caffaz, S., *Test di laboratorio per valutare la producibilità di biogas, Chapter 9*. In Vismara Renato, Malpei Francesca, Canziani Roberto and Piccinini Sergio (Eds.) *Biogas da agrozootecnica e agroindustria*, Dario Flaccovio Editore, Palermo, Italy, 2011.

Hafner, S., Richards, B.K., Astals, S., Holliger, C., Koch, K., and Weinrich, S., Calculation of methane production from gravimetric measurements, September 7 2020. https://github.com/sashahafner/BMP-methods/raw/master/203_gravimetric_calculations.pdf.

Hafner, S., Justesen, C.G., Thorsen, R., Astals, S., Holliger, C., Koch, K., and Weinrich, S., Calculation of methane production from gas density-based measurements, October 12, 2020. https://github.com/sashahafner/BMP-methods/raw/master/204_gas_density_calculations.pdf.

Hansen, T.L., Schmidta, J.E., Angelidaki, I., Marca, E., La Cour Jansen, J., Mosbæka, H., and Christensen, T.H., Method for determination of methane potentials of solid organic waste, *Waste Management*, *24*, 393–400, 2004.

ISO Norm 5725/1994 corr. Accuracy (trueness and precision) of measurement methods and results-Part 1: General principles and definitions, 1998.

Nistor, M., Strömberg, S., and Liu, J., Automatic methane potential test system II (AMPTS II) as a tool for screening and selecting organic substrates for anaerobic digestion processes, Proceedings of the Fourth International Symposium on Energy from Biomass and Waste, San Servolo, Venice, Italy, 2012.

Schnurer, A., and Jarvis, A., *Microbiological Handbook for Biogas Plants*, Swedish Waste Management U2009:03, Swedish Gas Centre Report 207, 2010.

Strömberg, S., Nistor, M., and Liu, J., Standardization as a key to reliable gas measurements in biochemical methane potential (BMP) tests, Proceedings of the Fourth International Symposium on Energy from Biomass and Waste, San Servolo, Venice, Italy, 2012.

Thomson, W. (Lord Kelvin), Lecture on *"Electrical Units of Measurement", Popular Lectures I*, 73, May 3, 1883.

VDI (Verein Deutscher Ingenieure), *Norm VDI 4630/2006-Fermentation of organic materials. Characterisation of the substrate, sampling, collection of material data, fermentation tests*. Beuth Verlag GmbH, Berlin, Germany, 2006.

UNI (Ente Italiano di Normazione), *UNI/TS 11703:2018 Metodo per la misura della produzione potenziale di metano da digestione anaerobica ad umido – Matrici in alimentazione (Italian norm on BMP test protocol, in force since 2018*, https://store.uni.com/uni-ts-11703-2018), 2018.

3

How to Perform Anaerobic Tests under Optimum Conditions

3.1 Scope

The previous chapter explained the most suitable instruments and tests used for managing biogas plants, i.e., those techniques that are useful for optimizing the whole anaerobic digestion (AD) process. This chapter provides a practical guide for conducting tests in the biogas plant, supplemented with some valuable tips and hints to help avoid the common mistakes made by beginners. The practical application of the results obtained with such tests will be explained in detail in Chapter 4.

3.2 Measuring the DM and VS

Two tests are fundamental in optimizing the AD process: the determination of the DM and the volatile solids (VS) of the plant's feedstock (basic information to define the strategies for optimizing the substrate) and the measure of both the VS of the mixture fed to the digester and the digestate exiting it (useful information for optimizing the process).

The determination of both DM and VS is straightforward to carry out, but their importance is fundamental to understanding the feedstock's quality and the efficiency of the AD process. The feedstock's quality is directly correlated to its DM content, because moisture does not contribute to the production of biogas but does increase the transport costs. A high moisture content sometimes favors the growth of molds, which produce mycotoxins that can inhibit the AD process. Indeed, the main ingredient for the production of penicillin and other antibiotics is some selected mold species. In countries where AD feedstock is a commodity derived from energy crops, like Italy and Germany, it is customary to define the price as a function of the DM content. Such a

DOI: 10.1201/9781003676201-3

criterion is reductive because two lots of the same feedstock may have identical DM content, but different methane potential. The reasons can be either a different degree of lignification of the biomass, or a higher ash content, or both. In practice, measuring the ash content is helpful because the difference between DM and ash, called VS, represents the *potentially degradable* fraction of the organic matter. It is then possible to obtain a rough estimate of its biochemical methane potential (BMP). The words "potentially degradable" are highlighted because, even if having identical VS content, two lots of the same substrate can have different BMP, depending on the lignin content, on the presence of eventual inhibitors (molds, pesticides, etc.), on the C/N ratio, on the particle size, or eventual pretreatments (thermal, chemical, or biological) that the substrate may have undergone. For all the said reasons, the following test sequence is necessary for a correct analysis of the substrate's quality:

> determination of the DM → determination of the ash content (and hence the VS) → determination of the BMP and degradation time

The second application of the VS measurement is assessing the digestion efficiency. Since the quantity of biogas produced by the AD is directly proportional to the VS content, the ratio between the VS of the feedstock and the VS of the digestate is an index of the conversion efficiency of the biomass into methane (and hence, an indicator of the good or bad operation of the plant).

The sequence of the necessary tests to analyze the plant's overall efficiency is the following:

> Determination of both feedstock and digestate's VS → calculation of the digestion efficiency → determination of the digestate's residual BMP (complementary test).

3.2.1 Measuring the DM with a Moisture Analyzer

This is the simplest test on any feedstock and requires just following the user's manual of the instrument. The sequence is as follows:

1. Place a dish on the scale and bring the reading to zero.
2. Load uniformly the dish with the quantity recommended by the instrument's manufacturer (usually 10 g).
3. Close the heating lid, select the program (if any), and press the start button.
4. The instrument will start the drying cycle and stop it automatically when it does not detect any weight variation within a preestablished time. On reaching the said point, it will store in its memory the initial and final weights and the resulting moisture content.

Although the moisture analyzer is a very simple instrument, it can easily lead the beginner into errors, wasting time, and hampering the quality of the results. The following list shows the most frequent errors when measuring DM and VS.

3.2.1.1 Confusing the "Moisture" and the "DM" Scales

By default, moisture analyzers measure moisture (i.e., the weight difference between wet and dry samples) in percentage. Most commercial models allow selecting different ways of expressing the result: Moisture in % (of the wet weight; WW), DM in % (of the WW), ATRO moisture, and ATRO dry matter. The word ATRO comes from the German *Absolut TROcken* (absolute DM), and is a conventional way of measuring the moisture in use by the timber industry. For our purposes, the useful value is the DM expressed as % of the WW; hence, it is necessary to program the instrument to display the said value. To avoid any doubt, it is important to summarize the meaning of each of the conventions defined previously:

1 Moisture in % of the WW:

$$M[\%] = \frac{\mathrm{WW} - \mathrm{DW}}{\mathrm{WW}}$$

2 DM in % of the WW (the useful value for our scopes):

$$\mathrm{DM}[\%] = \frac{\mathrm{DW}}{\mathrm{WW}}$$

3 ATRO moisture (% moisture referred to the DM):

$$M_{\mathrm{ATRO}}[\%] = \frac{\mathrm{WW} - \mathrm{DW}}{\mathrm{DW}}$$

4 DMATRO (% of WW referred to DM):

$$\mathrm{DM}_{\mathrm{ATRO}}[\%] = \frac{\mathrm{WW}}{\mathrm{DW}}$$

3.2.1.2 Wrong Setting of the Drying Temperature

As a rule, measuring the DM of feedstock for biogas production (typically silage, manure, etc.) requires a slow drying process, possibly setting the temperature at 105°C. A quick drying process, at high temperature, may bring the sample's water to boil, scattering small particles out of the sample dish, thus biasing the test's result. When measuring liquid substrates (e.g., sludge and digestate), the surface will dry first, forming a watertight layer that prevents the evaporation of the remaining moisture. To prevent the said phenomenon, adding a certain quantity of sand to the sample will help the heat conduction and break the surface tension of the liquid.

N.B.: The moisture analyzer measures just the weight difference, so the sand's additional weight poses no problem: just add it to the dish, press the tare button to bring the reading to zero, and then add the liquid sample to test.

3.2.1.3 Wrong Installation of the Moisture Analyzer

Like any other electronic instrument, the moisture analyzer's readings are subject to thermal drift, meaning that changes in room temperature can produce variations in the readings. It is of utmost importance to protect the moisture analyzer from air currents and direct sunlight, placing it away from heaters, radiators, windows, or air conditioners. It is a good practice to turn the instrument on at least 15 minutes before starting the measurement to stabilize its internal temperature. Sometimes, moisture analyzers include either a precision weight for manual calibration or a self-calibration function. It is highly advisable to calibrate the instrument, especially during extreme temperature seasons or after turning off the laboratory's heating or air conditioning (such as during the night or weekends). If you don't have a calibration weight, it is advisable to purchase one with adequate accuracy. The necessary weight and its accuracy, i.e., the class, are stated in the balance's manual. For example, the author's moisture analyzer (100 g maximum weight, ±1 mg accuracy) requires a Class F1 calibration weight (100 g ± 0.5 mg). Some upper-range balances may require two calibration points.

3.2.1.4 Measuring the DM of Silage

The ensiling or silaging process is a special type of fermentation, similar to the preparation of *Sauerkraut* (aka *choucroute* or *pickled cabbage*), which converts the sugars contained in the fresh vegetable matter into volatile fatty acids (VFA), primarily lactic and acetic acids. Some feedstocks, such as grape marc and silage of sugar beet leaves, also contain ethanol. If the silage is kept away from air, the acids act as a preservative, ensuring its nutritional properties. Lactic and acetic acid have boiling points close to that of water (122°C and 118°C, respectively). Hence, they evaporate when measuring the DM of silage. Since lactic acid is a direct precursor of methane, it is not correct to account for it as water when drying the biomass to obtain the DM, and successively the VS, because it is indeed VS itself. Since VS is the denominator in the BMP's formula, underestimating the VS translates into a wrongly high BMP value. The solution to avoid the evaporation of the lactic acid (and any

VFA in general) when drying the silage to measure its DM consists of adding a known amount of some hydroxide, usually caustic soda (NaOH) or spent lime [$Ca(OH)_2$]. This will form sodium or calcium lactate, a salt that does not evaporate when drying, but volatilizes at 550°C. The difference between the DW and the net ash weight (total ash weight minus the hydroxide added to the wet silage) is then the correct VS value. This technique does not work well with sugar beet and sweet sorghum silage, because the sugars contained in such biomass ferment into alcohol, which does not react with NaOH and will evaporate anyway. A more complete, but more complicated method, consists of distilling the silage in a still, kept at near-boiling-point temperature (105°C) until dry, then collecting the condensate and measuring both the dry mass that remained in the still and the chemical oxygen demand (COD; sum of all volatile compounds) collected in the condensate. The method was described by Porter and Murray (2001).

3.2.1.5 Interpretation of the Scale's Technical Sheet

The error margin of the DM determination is typically specified in the technical sheet or manual provided by the manufacturer of the moisture analyzer. It ranges from ±0.015% (for upper-range models) to ±0.25% (for low-cost models). Some manufacturers specify two error margins: one for 10-g samples (maximum accuracy) and another for 2-g samples (minimum test time, but lower accuracy). In this particular case, "%" is the measure unit of DM, so for a technical sheet reporting, for instance, the specification "±0.15%" actually states that the absolute error in the DM determination is ±0.15%. In this particular case, the sign "%" must not be confounded with the relative error (conventionally expressed as a percentage of the instrument's measure limit) or with the instrument's sensitivity (aka. resolution, i.e., the minimum difference in weight that the scale is capable of measuring).

Example: The technical sheet of a moisture analyzer specifies a maximum error of 0.05% when measuring moisture with 10-g wet samples. Suppose we test a 12-g sample of wet biomass, obtaining 32.53% DM. According to the technical specification sheet, the actual DM value could then be any value in the range of 32.48%–32.58%. The correct way to express the DM in this example would be 32.5%, because the centesimal figures are not relevant—we only know for sure that the DM value is bigger than 32% and smaller than 33%.

3.2.2 Measurement of the DM with an Oven and a Scale

Carrying out this test requires a lower cost of investment compared to the automatic moisture analyzer: it is enough to have a small electric oven or a hot air stove, setting the temperature at 105°C–110°C, and a scale with at least ±0.1 g accuracy, like the ones employed for weighing letters. Compared to the automatic measurement using a moisture analyzer, weighing the wet sample, then drying it in the oven, and weighing it periodically until no weight variation is detected, requires more manual labor and increases the possibility of human errors. In Chapter 2, Section 2.1.2, Example 1 demonstrates how to measure the DM with reasonable accuracy, even when using a domestic kitchen scale. From the said example, we can deduce that the lower the scale's accuracy, the bigger the sample size necessary to obtain an acceptable result; hence, the higher the energy cost and the longer the time required to carry out the test. According to the Standard Methods, it is necessary to weigh the sample at regular intervals until no variation in weight is observed. Since most biomasses are hygroscopic, the sample should be weighed while still hot (but the heat can be transmitted to the scale's weight sensor, altering the reading) or left to cool in a glass bell filled with silica gel to prevent it from absorbing moisture from the air. A more comfortable, but energetically more expensive method consists of leaving the sample in the oven overnight and checking the weight in the morning with short intervals. In general, a whole night in an electric oven is sufficient to dry a 200-g biomass or liquid sample, while 4 hours may be sufficient when employing a forced convection oven. Summarizing, the procedure to measure the DM of any feedstock is the following:

1. Calculate the sample quantity as a function of the scale's accuracy [see Chapter 2, Section 2.1.2 (Example 1) and Table 3.1 in the next paragraph of this chapter].
2. Place the sample in the oven, previously heated to 105°C, and leave it for a time long enough for its complete drying.
3. Weigh the sample while it is still hot, or leave it to cool in an airtight container filled with silica gel to prevent moisture absorption from the air.
4. Calculate the DM as % of the WW with the usual formula:

$$\mathrm{DM}[\%] = \frac{\mathrm{DW}}{\mathrm{WW}}$$

3.2.3 Measuring the Ash Content and VS

Measuring the ash content and consequently the VS requires inevitably a scale with the best resolution possible, because ashes are usually less than 10% of the WW, often a fraction ranging from 2% to 5%. The standard procedure to measure ashes is the following:

1. Determine the DM with any of the two methods described in Section 3.2.1 or 3.2.2.
2. Set the muffle oven at 550°C and wait until the temperature has stabilized (depending on the muffle's control fineness, in general,

when reaching 550°C, the temperature will cycle around the said value).

3 Place the dry sample in the hot muffle oven. Check that the smoke flows freely from the oven's fume hole or chimney.

4 Leave the sample at 550°C until complete calcination. Typically, 4 h is enough. The dish's bottom must be covered with white–gray ash (if char rests or embers are observed, place the sample back in the muffle oven for a few additional hours).

5 Weigh the sample while still hot or let it cool in an airtight container filled with silica gel, to avoid moisture absorption or hydrate formation.

6 Calculate the VS as follows:

$$VS[\%] = \frac{\text{DW} - \text{Ash}}{\text{WW}}$$

7 Check the error in VS determination, because the said value will be useful for the BMP test. The procedure to check the error propagation is the following:

Applying the theorem of the error of the quotient between two measured magnitudes to the definition of VS presented in point (f), we obtain the following:

$$e_{\text{vs}} = e_{(\text{DW}-\text{Ash})} + e_{\text{WW}}$$

The first term can be further developed as follows:

$$e_{(\text{DW}-\text{Ash})} = \frac{E_{\text{DW}} + E_{\text{Ash}}}{\text{DW} - \text{Ash}} = \frac{2E_{\text{scale}}}{\text{DW} - \text{Ash}}$$

The second term can be simply calculated with the following:

$$e_{\text{WW}} = \frac{E_{\text{scale}}}{\text{WW}}$$

Hence, the total error in the VS determination is

$$e_{\text{VS}} = \frac{2E_{\text{scale}}}{\text{DW} - \text{Ash}} + \frac{E_{\text{scale}}}{\text{WW}}$$

Practical Rule

The following practical rule is a consequence of these formulas.

In the case of "normal" substrates (silage, dung, vegetal waste, etc.) containing around 30% DM and 3% ash (referred to the WW), the sample size necessary to determine the VS with a given admissible maximum error (e.g., 1%) can be calculated as a function of the available scale's maximum error, with the following formulas:

$$e_{VS} = 0.01 = \frac{2E_{scale}}{0.3 \cdot WW - 0.03 \cdot WW} + \frac{E_{scale}}{WW}$$

$$0.01 = \frac{2E_{scale}}{(0.3 - 0.03) \cdot WW} + \frac{E_{scale}}{WW}$$

$$WW = \frac{2E_{scale}}{0.27 \cdot 0.01} + \frac{E_{scale}}{0.01}$$

$$WW = 740 \cdot E_{scale} + 100 \cdot E_{scale} = 840 \cdot E_{scale}$$

For instance, if the available scale has a maximum error of 0.1 g and we want to determine the VS of silage with a 1% maximum error, the sample size necessary to dry and successively to calcine will be at least 84 g. If a moisture analyzer with an absolute error of 0.005 g (a frequent value for medium-quality instruments) is employed instead, then a 4.2-g sample will be sufficient. Since, in general, the specification sheets of moisture analyzers require an initial sample size of 10 g, it is possible to conclude that the same sample, once dry, can be calcined, and the weight difference with the ash will represent the VS, without additional error.

Table 3.1, based on the practical rule deduced earlier, provides a useful guideline to calculate the quantity of sample necessary to measure the VS as a function of the absolute error of the available scale and the estimated moisture of the material under analysis.

TABLE 3.1

Quantity of Sample Necessary to Keep the Error of the Calculated VS below 1%, as a Function of the Absolute Error of the Scale, E_{scale}, and of the Estimated Moisture of the Substrate

Substrate under Analysis	Substrate's Approximate Moisture	Minimum Quantity of Sample to Keep $e_{vs} < 1\%$
Dry: straw, meals, oilseed cake	Up to 40%	$400 \cdot E_{scale}$
Semidry: silage, grass, municipal waste, fresh manure	40%–80%	$1000 \cdot E_{scale}$
Wet: digestate, sludge	90%–95%	$3900 \cdot E_{scale}$

3.3 Measuring the COD and Total N

There are two primary methods for conducting these types of tests: analytical methods and photometric methods. The first ones are not particularly complicated, but are labor-intensive and require clumsy tooling. On the other hand, they allow testing of both liquid and solid substrates.

The photometer is suitable for testing liquid substrates that do not contain macroscopic solids. Its use in industrial biogas plants is hence limited to substrates such as whey, sludge (filtered), olive mill wastewater, digestate, and similar ones.

Particular attention must be paid to the maximum value that the instrument is capable of measuring (aka top of scale). The top of the scale varies from one manufacturer to the other, and in some cases, even has three different scales in a single instrument—for instance, COD LR (low range, from 0 to 150 mg/L), COD MR (medium range, from 0 to 1500 mg/L), and COD HR (high range, from 0 to 15,000 mg/L). In the biogas industry, sludge and liquid substrates typically have COD concentrations exceeding 40,000 mg/L. To utilize the photometer, it will be necessary to dilute the substrate with distilled water, sample the required diluted amount using the automatic micropipette, measure the COD, and then multiply the instrument's reading by the dilution factor.

Practical Example

It is necessary to measure the COD of cheese whey with a photometer model HI 86814. According to the manufacturer, the error margin of the COD HR scale is ±220 mg/L.

The instrument's top of the scale is 15,000 mg/L. From the literature, it is known that the COD of whey is in the range of 40,000–80,000. The error of the burette employed to sample the whey is ±0.1 mL, and the error of the cylinder where the dilution will be made is ±0.5 mL. Define the test protocol and calculate the maximum error.

To facilitate the calculations and to obtain at least the order of magnitude of the sample's COD, it is advisable to perform a first test with a dilution factor equal to 10. In this way, the COD value of pure whey will result from multiplying the instrument's reading by 10. The whole procedure is as follows:

1. Dilute 10 mL of whey in 90 mL of distilled water, mix well, and sample the necessary quantity with the instrument's automatic micropipette (200 μL in this example).
2. Add the prescribed quantity to the vial, and follow the procedure defined by the manufacturer in the instrument's manual (usually incubate at 120°C for 2 h, mix gently, cool down, and finally measure with the instrument, but this may vary from model to model).

3 In our example, the measured value was 6,880 mg/L, hence

$$COD_{real} = Dilution \cdot COD_{measured} = 10 \cdot 6{,}880 = 68{,}800 \left[mg/L \right]$$

4 The error of this measure will be the sum of the single errors—the error of the photometer plus the error of the graduated cylinder employed to prepare the diluted sample. Hence,

$$e_{COD} = \frac{220 \left[mg/L \right]}{6880 \left[mg/L \right]} = 3.2\%$$

$$e_{dilution} = \frac{0.1\,mL}{10\,mL} + \frac{0.5\,mL}{100\,mL} = 1.5\%$$

$$e_{total} = 3.2\% + 1.5\% = 4.7\% \approx 5\%$$

5 Now that we know with good approximation the COD of the substrate in question, two alternatives are possible: to repeat the test with a dilution factor equal to 7, to obtain a measured value nearer to the real value, but with a smaller error margin, or keep this measure as the good one, accepting ±5% uncertainty.

If the liquid substrate under test contains macroscopic and suspended solids, like in the case of liquid manure or digestate, the procedure becomes more complex. Firstly, it is necessary to take a sample that can be assumed to be "representative" of the whole lot (for instance, by thoroughly mixing before taking the sample, in such a way that the liquid is as homogeneous as possible). It is necessary to filter the sample through a sieve (a stainless steel or plastic kitchen sieve, similar to those employed for sieving flour, will suffice). The solids separated in this way (suspended solids) must be calcined in the muffle oven to determine their VS. The liquid fraction must be diluted and its COD measured as already described in the earlier example. The total COD will be, with good approximation, the result of the following formula:

$$COD_{tot} = COD_{l.f.} + 1.3333 \cdot VS_{s.f.}$$

The COD calculated in this way tends to be slightly underestimated.

It should be decided case by case (depending on the solids content of the liquid substrate) if it is convenient to measure the COD as described, or if it is easier to evaporate the sample until completely dry and then to determine the VS. If the liquid contains alcohols or VFA as well as solids, then measuring only the VS is insufficient because the total organic matter content would be underestimated.

Measuring N (total and/or ammonia) with a photometer is analogous to measuring COD. As a general procedure, it is necessary to check the measurement range of the instrument, then find out the probable N content from the literature, and finally decide whether to dilute the sample or not. The eventual presence of solids is not a problem, because in most cases, more than 90% of the N remains in the liquid fraction. It is furthermore possible to wash the filtered solid fraction with a given amount of distilled water and then measure its N content, as nitrates and ammonia are soluble; hence, the total error will be further reduced. For routine controls, the method described is acceptable. Should it be necessary to determine the N content in nitrogen-rich semisolid substrates (such as chicken dung and swine manure), it is advisable to take a sample, test half of it with our instruments, and send the other half to a specialized laboratory. With the cost of a single test, we will then assess the accuracy of our field method, allowing us to correct our future measurements, as the difference between our values and those obtained by the specialized laboratory can be considered a systematic error. In general, our field measurements will be underestimated because the photometer cannot measure the nitrogen (N) contained in the solid fraction. Hence, comparing two tests made on the same sample, one in the field using the approximate method and the other performed by a specialized laboratory (assumed to be errorless), will provide us with the correction factor to apply to future tests.

3.4 How to Design a Biological Experiment

The scope of biological tests is measuring the quantity and chemical composition of the biogas (or the net quantity of methane) produced by a given quantity of substrate, under predefined reference conditions, by actually digesting the substrate with a suitable inoculum.

In this section, we will outline the necessary data and procedures required to conduct a biological experiment. It is essential to obtain the following information before launching any biological test:

1 The VS concentration (or the COD) of the substrate that must be tested. The said value is an indicator of the presumably digestible fraction. In general, it is possible to have an approximate idea of the order of magnitude of the substrate's BMP (at least for usual substrates); hence, based on the estimated yield, it is possible to define the amount of substrate to be employed in the test. For instance, suppose that the available instrument is an eudiometer having a 250 mL capacity. Suppose that corn silage must be tested, which presumably will produce between 350 and 500 mL/g SV. Hence, the maximum amount of VS

that can be employed is 0.5 g; otherwise, it will be necessary to vent some of the biogas periodically during the entire test period, typically 30 days, which in turn will increase the total error. When using electronic instruments like the AMPTS III, there is no problem because they have enough memory capacity to collect data for several months.

2 *The VS concentration of the inoculum*: This datum provides an idea of the quantity of active microorganisms capable of degrading the substrate. The VS concentration of the inoculum is of fundamental importance, since it will directly influence the error propagation.

3 The technical features of both the available reactors and the gas measurement system. It is necessary to have a clear idea of the reactors' volumes, the error margin of the gas volume measurement device, and the measurement range (of flow, volume, or pressure of the produced gas). It is essential to define whether the device is measuring gross biogas or net methane production. In the first case, the error of the gas composition analyzer will be added to the other instrumental errors.

4 *The scope of the test*: Do we need to check the health state of our plant, or the BMP of a given organic matter? We must define clearly what information we need to obtain.

5 *The representativeness of the sample*: Some substrates (e.g., municipal organic waste) are extremely heterogeneous. It is impossible to sample a full truck of municipal waste, homogenize its content, and extract the few grams of homogeneous sample required for our tests from the resulting mass. In such cases, there are two alternatives to measure the BMP:

   a To employ a big reactor or pilot plant, capable of receiving many kilograms of feedstock each day (assuming that the said amount is sampled randomly in such a way that it can be considered "representative"), or

   b To perform a large number of independent laboratory tests, whose results will be analyzed in statistical terms (average and dispersion of the BMP values, seasonal trends, etc.). With the same criterion, if the scope of the test is measuring the BMP of liquid manure from a stall, it will be necessary to mix thoroughly the content of the stall's catch basin, to avoid sampling just the surface scum, or the sediment, and to obtain a sample as homogeneous as possible.

Once all the above questions are clear, it will be possible to start preparing for the test, regardless of which test it is. The preparation phases, which are common to all anaerobic biological tests, will be explained in the following paragraphs. The descriptions are, in general, valid for any instrument, unless otherwise stated.

3.5 The Preparation of Both Inoculum and Sample

As a rule, the inoculum for our tests will always be a certain amount of sludge taken from the digester (in the case of single-stage biogas plants). If the AD plant consists of several stages, the inoculum must be sampled from different tanks, depending on the scope of the test:

1. The inoculum for the SMA test (described in Section 3.10.5) must be sampled from the primary digester.
2. The inoculum for testing the fermentative activity must be sampled from the prefermenter or the first stage.
3. If the scope is to measure the BMP of a given substrate, it is possible either to employ a mixture of sludge samples taken from each reactor or to employ digestate extracted from the last stage. Digestate has a smaller residual BMP than active sludge, so it requires a shorter preincubation time, and the uncertainty of the result is lower because of the reduced endogenous gas production. The influence of the "background noise" introduced by the inoculum's endogenous gas production will be explained in detail in the next chapter.
4. To measure the residual BMP (i.e., the conversion efficiency of the plant), a sample must be taken from the digestate tank or post-fermenter.

Depending on the literature consulted, it is possible to appreciate that two different schools of thought exist about how to prepare the inoculum. The first one, with roots in the wastewater treatment industry, states that the inoculum should be taken from a plant that is already in stable operation, washed, centrifuged, and then resuspended in water. Table 3.2 shows the norms that adopt the said criterion. The resulting VS concentration is usually between 1 and 5 g/L.

The second school of thought, with roots in the agricultural and agro-industrial biogas industry, focuses on the potential production of energy through AD and not on the degradation of a given pollutant. Nevertheless, for the said scope, there is no universally recognized standard, although several have been proposed, such as the German norm VDI 4630 *Fermentation of organic materials—Characterization of the substrate, sampling, collection of material data, and fermentation tests*. The current version, as of July 2025, is VDI 4630:2016-11, available in both German and English. The said norm provides guidelines for tests aimed at optimizing biogas plants and assessing the greenhouse gas potential of waste for environmental studies. Still, it does not consider error propagation issues, limiting the analysis to evaluating the maximum dispersion between samples. The inoculum should be sieved through a 10-mm mesh and preincubated to reduce its endogenous methane production. No medium or any other substance that can be considered a

TABLE 3.2

International Norms on the Anaerobic Degradability of Some Organic Pollutants, Most of Them Applicable to Wastewater Treatment

Body-Nr. (Year)	Title
ECETOC no. 28 (1988)	Guideline for screening of chemicals for anaerobic biodegradability
ISO 11734 (1995) EN ISO 11734 (1998) BS EN ISO 11734 (1999) UNI EN ISO 11734:2004	Water Quality—Evaluation of the ultimate anaerobic biodegradability of organic compounds in digested sludge—Method by measuring biogas production
ASTM E2170 (2001)	Standard test method for determining anaerobic biodegradation potential of organic chemicals under methanogenic conditions
OECD 311 (2006)	Anaerobic biodegradability of organic compounds in digested sludge by measurement of gas production
ASTM D5210—92 (2007)	Standard test method for determining the anaerobic biodegradation of plastic materials in the presence of municipal sewage sludge
ISO 15985:2004	Plastics—Evaluation of the ultimate anaerobic degradation in conditions of high solids content AD— Method based on the analysis of the produced biogas.
ISO 15473:2002	Soil quality—Guidance on laboratory testing for biodegradation of organic chemicals in soil under anaerobic conditions

"booster" must be added, except for the case of liquid single-substrate fermentation. The inoculum's biological activity should be checked employing reference substrates (sodium acetate and cellulose). Many operators consider that the procedures required by VDI 4630/2014 are too complex for industrial plant laboratories to comply with.

Another proposal, cited at the end of this chapter, was presented by Angelidaki et al. at a congress of the *International Waste Association* (IWA). According to the authors (in line with many other researchers), inocula taken from agricultural AD plants (energy crops or agro-industrial byproducts) *should not* be washed and resuspended. Any pretreatment on the inoculum should be limited to separate eventual inert matter (wood chips, silt, small stones), but the rest of the vegetal fiber must not be eliminated, because it serves as a support for the microorganisms to grow; hence, they are useful to maintain the biodiversity of the bacterial ecosystem. We will see later on that such a preconception can lead to increased error propagation.

The author was involved in the development of the Italian standard UNI/TS 11703:2018, focused on the BMP batch assay for industrial AD. Testing the greenhouse gas potential of biomass is out of the norm's scope. The norm provides two protocols, one "complete" (similar to the VDI 4630) and a "simplified" assay, consisting of a blank reactor filled with only inoculum and two sample replicates. The inoculum should be sieved with a 5-mm to 10-mm mesh, and its VS content must be less than 5%; otherwise, it should be diluted. Medium must not be added to each reactor unless one is aware of the eventual shortage of microelements. Adding nutrients to an inoculum that already has a sufficient concentration may bias the result, sometimes lowering the BMP. It is admissible to pretreat the substrate in any way, but such pretreatment must be clearly stated in the test report. The inoculum must be preincubated until the daily production reaches 10% of the total cumulative volume, and its bacterial activity should be checked periodically (at least three or four times a year), using reference substrates such as sodium acetate and cellulose. The standard includes an informative chapter on error propagation, but it is not mandatory to include the error analysis in the test report. This is in contrast with the ISO/IEC Directives, Part 2: *Principles and rules for the structure and drafting of ISO and IEC documents*. The said directives state that a norm on laboratory test methods must establish the acceptable instruments, their calibration protocol, the estimation of the test's accuracy, and even the rules for rounding the results.

It is thus clear that neither the German VDI 4630 nor the Italian UNI 11703 norms comply with the ISO directives, so there is still some room for improvement in the accuracy of anaerobic biological tests.

The main features and flaws of the VDI 4630, the IWA protocol, and the UNI/TS 11703 standards are detailed in Chapter 6.

An important aspect that the AD plant manager must consider is the *specificity* of the inoculum. Inocula employed by university laboratories are typically prepared from sludge and digestate from multiple plants, resulting in highly biodiverse bacterial ecosystems that can quickly adapt to various substrates. Sometimes, but not always, it is observed that BMP values published in scientific congresses and papers tend to be higher than those effectively encountered during the operation of AD plants or in one's own laboratory. One of the many reasons is the inoculum's specificity or ability to digest a given substrate. A typical situation encountered in practice is that of agricultural biogas plants. Figure 3.1 shows the result of a real experiment on pure cellulose, using two different inocula. The first one was sludge from an anaerobic sewage treatment plant, as required by the German VDI 4630. In contrast, the second was sludge from an agricultural digester usually fed with corn silage and bovine manure, which is the second choice according to the said standard.

It is then clear that the bacterial ecosystem of the sludge from an agricultural biogas plant is quite well-adapted to digest short-chain polysaccharides and cellulose. Suppose that in a given moment, the plant's manager has the possibility of getting free, or at very low cost, some feedstock containing proteins and fats, for instance, slaughterhouse waste or residual glycerol from the biodiesel industry. According to scientific literature, such a feedstock has a high BMP, but in the author's experience, feeding it directly to an agricultural digester may cause some trouble. Since the bacterial ecosystem of cow manure is not adapted to degrade high concentrations of protein and fats, the most probable effect will be to slow down the plant's productivity and,

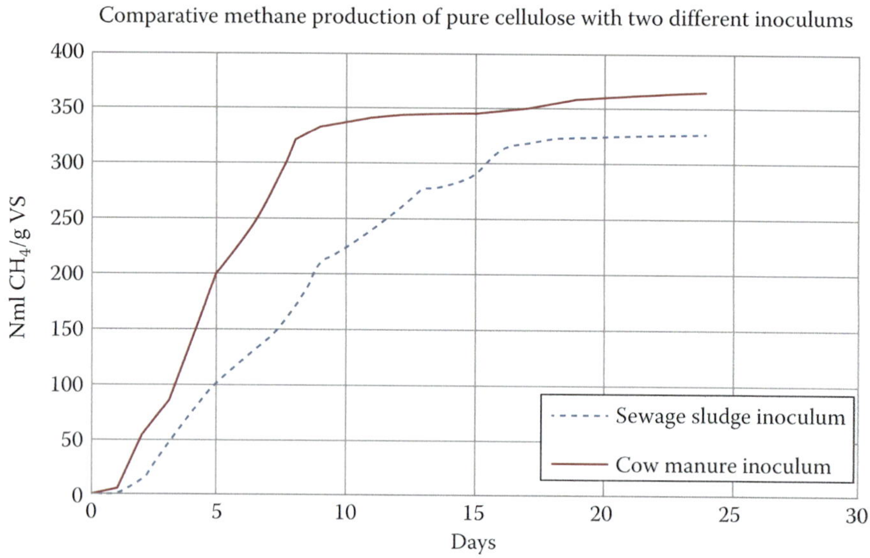

FIGURE 3.1
Digestion of pure cellulose with two different inocula. Since ruminants have specific cellulolytic bacteria in their guts, the inoculum from the biogas plant fed with bovine manure shows quicker and complete degradation of the substrate. Humans (and monogastric animals in general) do not possess such bacteria, so the degradation of cellulose with sludge from the sewage treatment plant was slower and incomplete.

in some extreme cases, even bring it to biological collapse. Sampling sludge from the digester and testing it in the lab, for instance, with the slaughterhouse waste, will produce a given BMP. Repeating the test with the same substrate, using the digestate from the first test as inoculum, will most probably yield a higher BMP. The reason is that the trophic relationships between the different species composing the bacterial flora will evolve and adapt to the new substrate, but once adapted (usually after two or more tests), the inoculum will be fit to digest both cellulosic and protein-based substrates. The shape of the degradation curve, which should fit more or less to the classical sigmoid, will indicate to the experienced user whether an inoculum is suitable for digesting a given substrate or not.

Figure 3.2 illustrates the digestion of a single sample of soapy residue from sunflower oil refining, which was tested with inocula from two different sources. Inoculum A was a mixture of inocula from different anaerobic digestion plants, but none of these had ever digested soapy waste. Inoculum B was sampled from a digester that is only fed with a mixture of soapy residues from refining palm, corn, and sunflower oils. Despite Inoculum A having an acceptable methanogenic activity (curve Acetate+Inoculum A), it was not able to digest the soapy residue, as we can deduce from the shape of the corresponding curve. On the other hand, Inoculum B had a lower methanogenic capacity (it

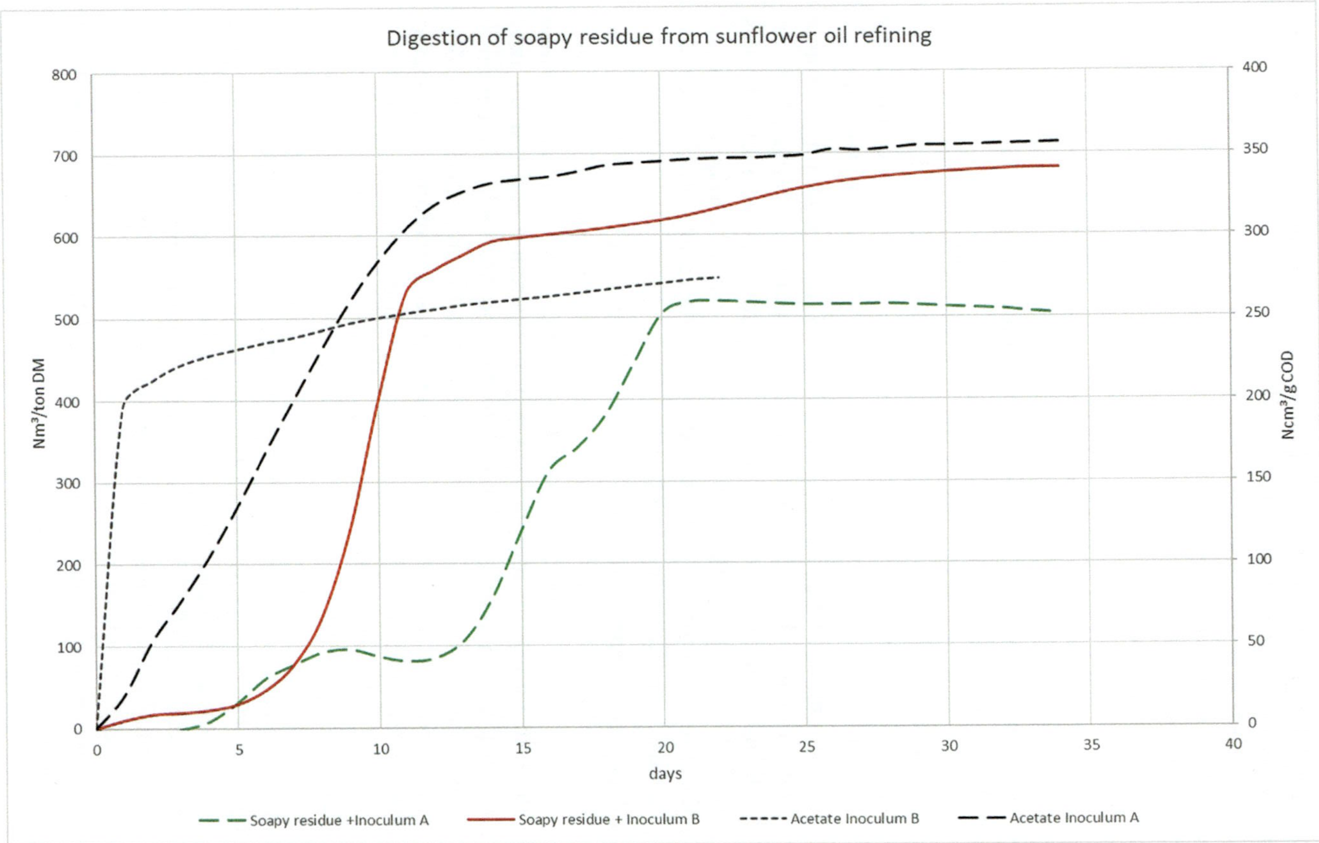

FIGURE 3.2
A single sample of soapy residue from sunflower oil refining yielded two different BMPs, depending on the inoculum employed for the test.

was unable to reach the expected BMP of sodium acetate, probably because of partial inhibition caused by the sodium added with the acetate, but could digest the soapy residue, albeit with a long lag phase.

At this point, it seems clear that, if planning to change the diet of the digester, it is the plant manager's responsibility to decide the test's scope and how to prepare the inoculum. For instance, if the manager plans to feed the biogas plant with a substrate having a very different chemical composition from that of the usual feedstock, it is necessary to check whether the bacterial ecosystem is capable of digesting it. It may be necessary to determine a strategy to adapt the existing inoculum to digest the new substrate.

The following are some of the possible strategies to apply when planning to include a new feedstock in the digester's diet:

1 Obtain digestate considered as "optimum" from a plant that is already regularly fed with the substrate in question, and run a comparative test with inoculum from your plant (difficult but not impossible).

2 If planning to restart a plant after an inactivity period, mix a sample of its sludge with sludge from the "donor" plant (the active plant that will supply the inoculum for the restart of the inactive one). Mixtures with different proportions of both inocula can help to determine the minimum amount of external inoculum necessary for a successful restart, minimizing transport costs.

3 Adapt the plant's inoculum to digest the new substrate. This is the easiest choice, but it may require a longer time if the initial biodiversity of the inoculum is too poor. In such a case, it is advisable to take the inoculum from the plant already in operation and test the new envisaged substrate mixed in different proportions with the currently employed feedstock. Eventually, the test can be repeated employing the digestate resulting from the first test as inoculum for a second test with an increased proportion of new substrate. The goal of this method is to find out the maximum amount of new substrate that the inoculum can digest without losing its methanogenic activity.

Finally, it is necessary to consider the preincubation of the inoculum before starting the test. The said operation, called *degassing*, is mandatory to exhaust the degradable organic matter already present when sampling the inoculum, so that the measured VS represents, with good approximation, the mass of living bacteria. Furthermore, a degassed inoculum will produce less *background noise* (the volume of gas produced by the inoculum itself). Degassing the inoculum sensibly reduces the error of the BMP assay and correlated tests. To degas the inoculum, follow the procedure described here:

1 Sample sludge from the digester, sieve it with a 5-mm mesh, measure its VS, and incubate it at the temperature chosen for the test (usually the same temperature at which the plant is running).

2. Observe the daily gas production. The inoculum is "degassed" when the cumulated production curve tends to remain flat for at least a few days, or when the daily gas production is less than a certain value, which varies according to different norms and protocols. The author considers that a good criterion to establish whether the inoculum is already degassed is to check that the daily net methane production is less than 1% of the total cumulated volume since the beginning of the degassing procedure, and remains so for at least 3 days. Another criterion is to check that the specific daily production (the total methane production in 1 day divided by the total VS in the reactor) is less than 5 Ncm^3/g VS.
3. As indicative values, the degassing procedure may take from 1 week in the case of sludge from a sewage treatment plant to 30 days when fresh cow manure or sludge from a plant running on energy crops is employed.

3.6 The Inoculum/Substrate Ratio, I/S

When performing biological tests, it is important to define the ratio between the "live weight" of the microorganisms and the "available food" for them. An acceptable measure of the live weight is the VS concentration of the inoculum after sieving and degassing it, as already explained in Section 3.5. The VS of the substrate provides a good measure of the "presumably biodegradable" organic matter in it. The typical ratio between the inoculum's VS and the substrate's VS for batch tests ranges from 1.6 to 3, to ensure that the quantity of living bacteria will be enough to consume all the substrate. Values smaller than 2 are not advisable, since they may sometimes lead to the phenomenon known as *inhibition from substrate* (in practice, bacteria can suffer from indigestion because of an excess of food, or because the said food is too heavy to digest). I/S ratio values higher than 3 present the risk that the amount of gas produced by the substrate is of the same order of magnitude as the inoculum's background gas production. The said situation is likely to arise if the inoculum is not completely degassed and/or if the substrate is hardly degradable (e.g., animal feathers and hair), and the result will be an amplification of the instrumental error of the whole test.

Convention: In the scientific literature, it is frequent to find the inverse definition, S/I ratio, and consequently, the resulting ratio values will be smaller than 1. Throughout this text, the conventional definition adopted by the author has been I/S, hence practical values ranging from 2 to 3.

Practical rule: Always perform your tests with I/S = 3 to stay on the safe side. Substrates suspected of containing inhibitory substances (e.g., industrial wastewater, residual glycerol from biodiesel production) may be tested with I/S = 5 to avoid inhibition.

Once the VS of both inoculum and substrate are known, it is necessary to calculate how much of each to load into the lab reactors. For this calculation, it is usually assumed that the densities of both inoculum and substrate are equal to 1 (in general, fairly true for the inoculum, but not always true for the substrate). The said convention simplifies the calculation and induces a negligible error.

The following example shows the calculation procedure.

It is intended to calculate the BMP of a substrate having 32% VS (on WW). The inoculum that has been taken from the digester and, once degassed and sieved, contains 7% of VS on WW.

Suppose that the available reactors are 500-mL bottles, and the instrument will be a volumetric measurement device. It is necessary to leave a reasonable space to collect the gas, called "headspace". If the test is to be carried out with a barometric method, the said headspace will typically range between 50% and 60% of the bottle's volume. If a volumetric method is employed, the headspace should range between 15% and 20% of the total volume. Since in our example we will employ a volumetric measurement system, the headspace of the reactor is arbitrarily chosen to be 20% of 500 mL. Consequently, the total volume occupied by the mixture (inoculum + substrate) will be 400 mL. Based on the considerations already exposed in the former paragraphs, and on the practical rule, the I/S ratio adopted for this test will be equal to 3. It is hence possible to express mathematically the test conditions with the following formula:

$$\frac{Q_{\text{inoc}}}{Q_{\text{subst}}} = 3 = \frac{V_{\text{inoc}} \cdot VS_{\text{inoc}}}{V_{\text{subst}} \cdot VS_{\text{subst}}}$$

Furthermore

$$V_{\text{inoc}} + V_{\text{subst}} = 400 \text{ mL}$$

It is hence necessary to solve a system of two equations with two variables (i.e., the volumes of inoculum and substrate, respectively). Replacing V_{subst} in the first equation by its equivalent, deduced from the second one, we obtain

$$3 = \frac{V_{\text{inoc}} \cdot VS_{\text{inoc}}}{(400 \text{ mL} - V_{\text{inoc}}) \cdot VS_{\text{subst}}}$$

$$3 \cdot (400 \text{ mL} - V_{\text{inoc}}) \cdot VS_{\text{subst}} = V_{\text{inoc}} \cdot VS_{\text{inoc}}$$

$$3 \cdot 400 \text{ mL} \cdot \text{VS}_{\text{subst}} = \text{V}_{\text{inoc}} \cdot VS_{\text{inoc}} + 3 \cdot \text{V}_{\text{inoc}} \cdot VS_{\text{subst}}$$

$$\frac{3 \cdot 400 \text{mL} \cdot VS_{\text{subst}}}{VS_{\text{inoc}} + 3 \cdot VS_{\text{subst}}} = V_{\text{inoc}}$$

By replacing the VS values, we obtain

$$\frac{3 \cdot 400 \text{ mL} \cdot 0.32}{0.07 + 3 \cdot 0.32} = V_{\text{inoc}} = 373 \text{ mL}$$

Even if it is not fully true that the density of both substrate and inoculum is equal to 1, the said assumption is fairly approximate for the inoculum and fluid substrates. Since the I/S ratio is measured in g, pouring 373 mL of inoculum on 27 g of substrate will produce a mixture with the desired I/S ratio, and its volume will be very close to 400 mL.

3.7 Defining the Mix Ratio (Intensity of Stirring)

Stirring is a parameter of fundamental importance for most industrial digesters. Too much stirring stresses the microorganisms, and too little stirring makes it difficult for the bacteria to access the nutrients. The calculation method for correctly setting the ideal intensity of stirring given the plant's size will be explained in Chapter 4. As far as laboratory reactors are concerned, the only easy thing to do is to procure by any means that they have at least a minimum of stirring. It has already been shown in the former chapter that many of the methods for measuring the gas amount produced by AD do not foresee stirring, mainly because of constructive reasons, but mostly for historical tradition, because most of them were initially conceived for testing wastewaters. When testing substrates with low solids content (<5% VS), and if the said solids are smaller than 1 mm, the difference between results obtained with static reactors and those obtained with stirred reactors will be negligible. The explanation is that the convective motion caused by the reactor's heating, together with the nutrients being in dissolved phase or in suspension, allows the bacteria to assimilate the nutrients at any time easily. In the case of agricultural or municipal waste plants, where the sludge usually contains from 8% to 12% VS (up to 20% in some special digesters, known as "semidry"), it is necessary to induce some mechanical stirring which must be also applied to laboratory reactors working in the same conditions. Some laboratory instrument manufacturers propose magnetic rod stirrers, which are just magnets coated with plastic, placed in the bottom of the reactor, which in turn is placed on a plate (sometimes equipped with an electric heater too) that generates a rotating magnetic field, forcing the magnetic bar to spin inside the reactor. Such a solution, once again, is borrowed from the industry of wastewater analysis, but is not suitable for testing agricultural inocula and substrates, because they are quite viscous and contain fibrous solids that hamper the magnetic bar's ability to rotate. Reactors having an electric motor and a gearbox connected to a shaft with paddles, or a bent rod, or helical propellers, are more efficient. Tests carried out with such reactors will be more accurate and their results closer to real-life plant operation.

Figures 3.3 and 3.4 show a simple batch reactor with a 1-L capacity, brushless motor, and a stirring rod, and a "professional" model with a 5-L capacity for continuous tests.

Laboratory operators and many scientific publications also adopt the stirrer's speed as the main criterion for comparisons between tests, but just because most laboratory stirrers have a revolutions per minute or an "RPM" scale. For AD plant optimization, the concept "stirrer's speed" is not

FIGURE 3.3
Two-liter semibatch reactor made with a modified ISO GL45 glass bottle, with brushless motor driving a bent rod. (Photo by the Author.)

FIGURE 3.4
Five-liter continuous reactor made of stainless steel with mechanical stirring provided by a brushless motor and built-in electronic drive. (Photo Courtesy of Bioprocess Control AB.)

useful at all. In other words, adopting a single RPM value for all tests means defining an internal standard of the laboratory arbitrarily, making it difficult to extrapolate the test results to a different instrument with a different kind of stirrer, not to mention the real-scale digester itself. The stirrer's speed is an inadequate criterion to define the necessary stirring intensity for a given reactor because the stirring intensity is not just a function of the speed, but also of the stirrer's shape, dimensions, Reynolds number at which it is operating, and finally of the reactor's geometry. It may sound strange to many readers, but the digester's temperature plays an important role too, because the viscosity of the sludge depends, among other factors, on the temperature.

The *Unit Power* is a very interesting criterion for defining a reactor's stirring intensity, because it is based on more rational parameters. It is easier to apply to electric stirrers, with or without gear reduction. This criterion, originally developed in the USA, consists in calculating the stirrer's set point based on easy-to-measure physical variables. The *Unit Power* is based on an empirical fact: the AD processes work well when the power at the shaft of the stirrer (regardless of whether it is a propeller or paddle wheel) is between 5 and 8 W/m³ of digestion volume. At the laboratory scale, this corresponds to 6–8 mW/L. It is not easy to measure the power at the shaft of a small motor with a gearbox or brushless technology, such as those typically employed for laboratory reactors. The correct way to measure mechanical power is to connect a torque transducer and a tachometer to the stirring shaft, then measure the torque and rotation speed, and calculate the power at the shaft with the reactor full of sludge as the product of torque by angular speed. Unfortunately, suitable instruments for measuring a minimal torque are difficult to find on the market, and in any case, it would be challenging to adapt them to existing laboratory reactors. Since such a test is sporadically needed in industrial anaerobic plants, buying a tachymeter and dynamometer is a superfluous expense. The method for adjusting the stirring intensity correctly in a real-size digester is explained in detail in the second volume, *Optimising Biogas Plants*.

At the laboratory scale, stirrers are either driven by DC motors with permanent magnets or by brushless motors, such as the one shown in Figure 3.3. *Only in the special case of DC motors,* it is possible to estimate, with good approximation, the mechanical power at the motor's (or gearbox's) shaft through purely electrical measurements. The procedure is the following:

1. With the motor disconnected from its power source, measure with a tester the resistance across its terminals, *R*.
2. Mount the stirrer on an empty reactor. Connect the motor to its DC power source and let it run idle. Measure with the tester the voltage at the motor's terminals and the current. The idle DC power, W_0 (W), is the product of the voltage, U_0 (V), by the current, I_0 (A). The value W_0 represents all the energy losses (both electrical and mechanical). In the case of small reactors (<2 L), in general, the voltage will be of the order of some V, but the current will be of the order of a few mA, hence the product will be the power expressed in mW. The power dissipated as heat in the windings, called W_{Ω}, is calculated as the product of the resistance R by the square of the current. Hence, the mechanical power when turning idle, called W_{Mo}, represents all the frictional losses, both of the motor and its gearbox. W_{Mo} is calculated as the difference between W_0 and W_{Ω}

$$W_{Mo} = W_0 - W_{\copyright} = U_0 \cdot I_0 - I_0^2 \cdot R$$

3. Now repeat the test with the same stirrer, but this time with the reactor full of sludge. Since sludge is a non-Newtonian fluid, the power absorbed by the motor will vary with time until reaching a steady state, depending on the temperature and apparent viscosity of the sludge, usually after some minutes. Hence, it is necessary to keep the motor

stirring the sludge for at least 2 or 3 minutes before measuring voltage and current. Thus, we will be sure of measuring the actual stirring power in steady state. If in doubt about when reaching the steady state, it is possible to connect the amperometer from the beginning and wait until the current value is reasonably stable. At this point, the current (I) flowing through the motor's windings in steady state dissipates power as heat in the resistance (R) of the windings. Assuming that the voltage drop along the connection cables and internal resistance of the DC power source is negligible, the voltage at the motor's terminals, U, should be the same as U_0 measured in step 1 (if it is not, clean well the connectors and, if possible, use thicker cables or a power source with bigger rated power). Under the present operation with the reactor full of sludge, the current, I, will be higher than the one measured with the empty reactor, I_0. We will call W_t the total electrical power absorbed by the motor under load, and W_R the power dissipated as heat by the windings' resistance, calculated as the product of the square of the current I and the measured resistance R:

$$W_t = U \cdot I$$

$$W_R = I^2 \cdot R$$

4 Calculate the mechanical power dissipated by the motor's (or gearbox's) shaft on the sludge, W_m, as the difference between the total power under load, minus the frictional losses measured under idle condition, minus the heat dissipated across the windings' resistance under load. Hence,

$$W_m = W_t - W_{Mo} - W_R$$

5 Finally, the quotient between the power W_m and the volume of sludge in the reactor, V, must be a value comprised between 0.005 and 0.008 kW/m^3 (or between 5 and 8 mW/L, which is the same). In general, the said quotient will be higher than the ideal values, so it will be necessary to define *ON/OFF* cycles in such a way that, on average, the mechanical power at the shaft of the stirrer lies within the specified interval. Translated into formulas

$$5[\mathrm{mW/L}] < \frac{W_m \cdot t_{ON}}{V \cdot (t_{ON} + t_{OFF})} < 8[\mathrm{mW/L}]$$

6 Choosing t_{ON} and t_{OFF} is arbitrary. Nevertheless, we should consider another factor that helps increase the useful life of the stirrer

motors: the startup current. At each start, the current across the motor's windings can be a peak of several times the nominal current. Such peak overheats the windings, but if t_{ON} is long enough, the heat will have enough time to dissipate. Very frequent starts and stops do not allow enough time for the heat generated by the start current to dissipate. Consequently, it is a good practice to assume $t_{ON} + t_{OFF}$ equal to 3600 s (1 h), to grant enough time for the motor to cool down before the next start.

EXAMPLE

Assume we desire to set the stirring rate of a reactor equipped with a DC stirrer motor. The digestion volume is 400 mL. With the empty reactor, we measure the following voltage and current:

$$V_0 = 8.8\left[V\right]$$

$$I_0 = 42\left[\text{mA}\right] = 0.042\left[\text{A}\right]$$

Furthermore, the resistance of the windings is

$$R = 19\left[\Omega\right]$$

Hence

$$W_0 = 370\left[\text{mW}\right] = 0.370\left[\text{W}\right]$$

$$W_{\Omega©} = I_0^2 \cdot R = 0.042^2 \cdot 19 = 0.0335\left[W\right]$$

so

$$W_{\text{Mo}} = 370 - 33.5 = 336.5\left[\text{mW}\right]$$

Now we fill the reactor with the sludge and leave the stirrer running at full load for at least 2 min, at the temperature chosen for the biological test. After the mentioned time, we measure the following voltage and current:

$$V_t = 8\left[\text{V}\right]$$

$$I_t = 71\ [\text{mA}] = 0.071[\text{A}]$$

Hence

$$W_t = 0.568[\text{W}] = 568[\text{mW}]$$

and

$$W_R = I^2 \cdot R = 0.071^2 \cdot 22.4 = 0.113[W] = 113[\text{mW}]$$

The power dissipated on the sludge (the *Unit Power*) is therefore

$$\text{UP} = \frac{(568 - 336.5 - 113)[\text{mW}]}{0.4[1]} = 296\,\text{mW/L}$$

The given value is 37 times bigger than the one prescribed for the *Unit Power*, so stirring 97 s every hour (or 48 s every half hour) is the minimum stirring acceptable. It is not convenient to exceed the calculated *Unit Power*, because increasing the time the motors remain on only increases their wear, but does not change the biogas production.

Remark 1

The former example, taken from a real case (AMPTS II with 5 V DC motors and 30:1 gearbox), allows to calculate the efficiency of the DC motor and gearbox, as follows:

$$\eta_{\text{motor+gear}} = \frac{W_t - W_{\text{Mo}} - W_R}{W_t} = 20.9\%$$

Such a low efficiency is typical in tiny DC motors with a high gear ratio, as the one considered.

Remark 2

The error of the UP calculated as described above is relatively high and will depend on the error class of the instrument employed. If a high-quality instrument is not available, it is advisable to perform several tests on different stirrer motors and calculate an average value.

In the case of brushless motors, estimating the power at the shaft through electrical measurements is more challenging. Brushless motors are bi- or tri-phase motors driven by pulsating currents, so measuring their electrical power with a simple multimeter is not possible. Since a brushless motor requires an inverter to run, and the latter is fed with DC, it is possible to measure V_0 and I_0 at the inverter's input and calculate W_0. Then repeat the measurement with the brushless motor stirring the sludge, and calculate W_t. The difference W_t–W_0 is approximately the power dissipated on the sludge. The rationale for this approximation is that the power of a tri-phase motor is a function of the frequency rather than the current (not wholly true, but an acceptable approximation in the case of small brushless motors fed via an inverter with high-frequency rectangular pulses), so the ohmic losses in the windings plus the inverter are constant, regardless of the mechanical load at the shaft.

3.8 The BMP Test: One General Procedure, Multiple Applications

Now that we know how to obtain the basic data (VS of the substrate and inoculum) and how to plan a biological test, we will study the BMP assay in detail. In brief, the scope of the BMP assay is to measure the net methane quantity produced by any combination of inoculum and substrate—or a mixture of substrates—under specified test conditions. The BMP assay has multiple applications:

- Checking the effective methane yield of any biomass, mimicking as closely as possible the operational conditions of our biogas plant (i.e., with the actual bacterial ecosystem, at the same temperature, and with the same nominal HRT of our biogas plant). The scope in this case is to define the digester's diet and/or to check the purchase price of the feedstock.
- Checking the residual methane quantity that our biogas plant's digestate can still produce. The scope of this test is to measure the efficiency of the feedstock's conversion into energy.
- Measuring the health state of the bacterial ecosystem (the procedure is explained in detail in Section 3.10).
- Testing different operational strategies (different temperatures, different stirring rates, different particle sizes of the substrate, and others that will be described in detail in Chapter 4). The scope is deciding the most suitable management strategy for the biogas plant.
- Testing pretreatments (e.g., adding enzymes to silage) or commercial additives. The scope is to evaluate the effectiveness of the pretreatment or additive and assess whether the eventual improvements offset their higher operational costs.

3.8.1 General Procedure for Measuring the BMP

1. Degas the inoculum as already explained in Section 3.5. If employing unsieved digestate as inoculum, this operation will take longer but will provide a measure of the biogas plant's conversion efficiency (see Chapter 4).
2. Sieve the inoculum (if you haven't done so before) and prepare the mixture of substrate and inoculum in the correct I/S ratio, as already described in Section 3.6.
3. Fill at least two (better three) reactors with the said mixture, close them hermetically, connect them to the gas measurement device, and finally purge the head volume with an inert gas to eliminate any trace of oxygen. These reactors are referred to as sample replicates or simply *replicates*.
4. Fill at least one reactor (better two, three may be superfluous) with only the degassed inoculum. The said reactor(s) is (are) called *blank*.
5. Adjust the stirring intensity according to the guidelines in Section 3.7.
6. Start the experiment. If employing manual reading instruments, take at least one reading per day, *always at the same time*.
7. Copy the reading in a spreadsheet. If employing instruments that do not have real-time normalization capability, then using a barometer and an ambient thermometer is mandatory. Copy the readings of ambient P and T into the spreadsheet, which will be necessary to normalize the gas volume reading using the formulas explained in Section 2.2.1.2 in Chapter 2.
8. If employing a CO_2 trap (a bottle filled with NaOH solution) between the reactor and the measuring device, the gas volume reading represents the net amount of methane. If employing a barometric method or a volumetric method without a CO_2 trap, it is necessary to measure the biogas composition using a suitable method. *Do not assume that biogas is always a mixture of 60% methane and 40% CO_2, because such a convention is seldom true.* Copy the measured biogas composition into the spreadsheet.
9. At the end of the experiment (conventionally 30 days), make a backup copy of the spreadsheet and elaborate the data as explained in Section 3.9.

3.9 Processing the Measured Data

At this point, we will have a table with "raw data". In the best case (an instrument such as AMPTS II or III, equipped with CO_2 traps, a data logger, and real-time normalization of the gas volume), we will have a table containing the normalized gas volumes with a temporal axis divided into constant intervals (typically hours or days). In the worst case, (eudiometer or similar instruments), we will have a table containing the daily volume of gas produced by the reactors, together with the ambient *P* and *T* readings. Finally, we need to assess the chemical composition of the gas, if CO_2 absorption flasks were not included.

3.9.1 Step-by-Step Data Processing

1 Start from a spreadsheet containing the following columns: day (0, 1, 2, 3, …, *N*), daily gas production (methane or biogas, according to the case) of the replicates no. 1, no. 2, etc., idem of the blank(s), ambient *P* and *T* at the moment of each reading.

2 In the first cell of a separate column, type the following formula: average of the replicates' readings minus average of the blanks, and such difference divided by the VS of substrate added to each reactor. Copy down until the last row, corresponding to day *N*.

3 For instruments lacking real-time normalization: in the column to the right of the former one, normalize the gas volumes using the formulas explained in Chapter 2, Section 2.2.1.1.

4 Plot the normalized values as a curve, employing the graph function of the spreadsheet.

5 Check the error propagation according to the procedure explained in Section 3.11. The said operation is not mandatory, but advisable, to evaluate the reliability of the calculated value. This is especially necessary when the results are doubtful (too low or too high) or when the replicates yielded gas amounts with more than 10% difference from each other. In the latter case, such differences could be the result of a hardware problem (reactors not completely airtight, problems with the measurement instrument), or may be the consequence of a human error during the preparation of the experiment (wrong quantities of substrate, substrate too heterogeneous or not enough homogenized, some contamination of the inoculum, etc.). The comparison between the error of the calculated average BMP and the dispersion between single replicates enables us to distinguish whether the uncertainty is metrological (accuracy) or procedural (precision), allowing for a more precise diagnosis of any potential problems or errors during the test. For example, suppose that the error of the calculated average BMP is in the order of 5%, but the dispersion of the single replicates is bigger than said value. In that case, the cause is most probably a human error in the preparation of the test, as will be explained in the following Section 3.9.2., or the substrate is too heterogeneous, some gas leak, or—more rarely—a problem with the instrument.

3.9.2 Frequent Causes of Problems during the BMP Test

3.9.2.1 Problems Caused by Human Errors

Carrying out BMP tests is easy and, when it becomes a routine, quite often the operator works on "automatic pilot". Distractions unavoidably lead to errors (of weighing, in the preliminary calculations for the experiment preparation and so on). Figure 3.5 shows an example from real life. It was planned to measure the BMP of sorghum and corn with an AMPTS II. For several reasons, it was necessary to start the test immediately, and there was no time to measure the VS *before* the test began. Hence, the mixture for the replicates was prepared assuming a VS from the literature, defining I/S = 4, to ensure that the actual ratio was at least 3. The test was started; the VS were measured the next day, finding 16% for sorghum and 24% for corn. In the case of the sorghum sample (dotted lines), the VS quantity effectively loaded in the reactors was 6.6 g (40 g WW), and the resulting I/S was equal to 5. In the case of the corn sample (continuous lines), the VS was 8.9 g (37 g WW), resulting in an I/S ratio of 3.7.

The instrument employed for this test, a second-generation AMPTS, is equipped with caustic soda CO_2 traps and normalizes the volume in real time. Therefore, the curves shown represent the normalized production of methane of each single reactor. Observe that the curves of both the blanks

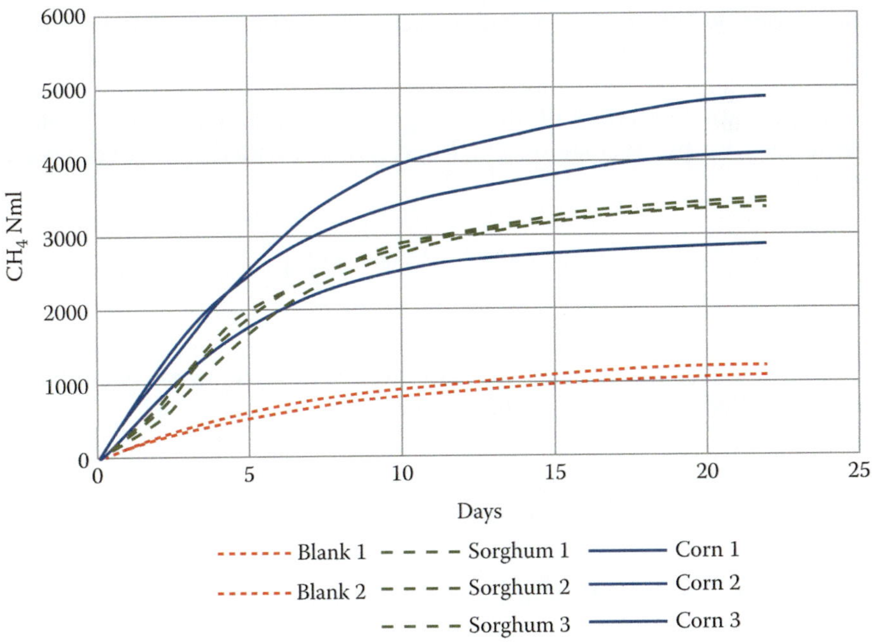

FIGURE 3.5
Curves of normalized methane production of each single reactor.

and the sorghum are almost coincident, while the curves of the corn samples are spread apart. Since the particle size of the corn sample was somewhat coarse, a certain degree of dispersion was expected. Such uncertainty is related to the representativeness of the sample rather than to the instruments or the method, but such a large dispersion was not expected. Observe that the curves *corn 1* and *corn 2* are almost coincident until the 5th day. Hence, their small dispersion can be attributed to the nonuniformity of the sample and its coarse trituration. The curve *corn 3* is suspiciously flat, although it follows the same shape as the other two. If a problem had been caused by the reactor (e.g., gas losses), or by the instrument (defects or issues in the sensor of the measurement cell, obstruction of the gas tube, etc.), the curve would be either interrupted or its shape would show "steps" or other forms of discontinuity. In this case, it is reasonable to suspect that the discrepancies between the curves can be attributed to human error. The test was prepared in a hurry, with the scope of teaching some students, so, most probably, the reason for the anomalous curve *corn 3* is simply that there was some mistake in the preparation of the sample. This example is helpful to justify why employing triplicate samples is always a good practice. Some protocols, such as the German VDI 4630:2016, specify that the number of replicates must be at least three. The Italian standard UNI/TS 11703/2018 considers at least two replicates as acceptable, to reduce the overall cost of each test. Observe that if the Italian protocol had been applied in the preparation of this experiment, such a considerable dispersion between curves would have led to discarding the test and repeating it. Since triplicate samples were employed and two of them gave coherent results, one can just discard the results of sample *corn 3* and calculate the BMP with the average of the remaining ones. Therefore, the result with the average of *corn 1* and *corn 2*, divided by the VS in each reactor, is

$$\text{BMP}_{\text{corn}} = \frac{0.5 \cdot \left[(4103 + 4881) - (1238 + 1112)\right]}{8.9} = 373\, Ncm^3/\text{g VS}$$

As a mental exercise, consider the said BMP as the base to calculate the VS that were effectively loaded in the reactor *corn 3*:

$$VS_{\text{presumed}} = \frac{2873\, Ncm^3}{373\, Ncm^3/\text{g VS}} = 2.4\,\text{g}$$

As can be observed, the WW of a sample containing such an amount of VS is 9 g. This means that the most probable cause of the anomalous curve *corn 3* is that the student mistook the quantity of VS with that of fresh matter to be loaded in the reactor. A typical example of human error.

3.9.2.2 Problems Caused by Instrumental or Method Errors

Sometimes, a reactor may be loosely closed, or a tube may be accidentally disconnected, causing gas leaks or even the entry of air, which can lead to a blockage in the process. Such is the case depicted in Figure 3.6, which is the result of some tests for measuring the BMP of *Arundo donax* (common cane, Castilian cane).

The reactor *Arundo2* had to be moved during the test, so it is likely that one of its tubes was accidentally loosened, allowing biogas to leak and air to enter. In this case, the problem is easy to solve because the three curves were almost coincident until the reactor was moved on the 18th day of the test. We can discard *Arundo2* and calculate the BMP_{30} considering the average of reactors *Arundo1* and *Arundo3*.

Practical Conclusion No. 1

It is advisable always to perform the tests in triplicate. Instrumental problems can always happen, and even the most experienced laboratory operators, working with top-quality instruments, are not exempt.

Practical Conclusion No. 2

Once you start a test, do not move the reactors or the measuring device until your experiment is finished, to avoid gas leaks.

We will conclude this paragraph with an example of the errors originating from the work method. An undergraduate student was assigned the task of

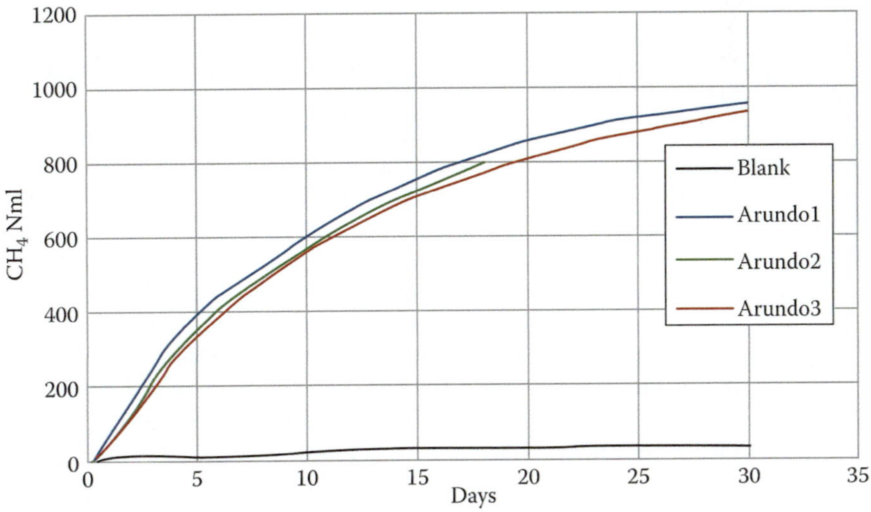

FIGURE 3.6
Curves showing the normalized production of methane from fresh *A. donax*. Observe the interruption of the AD process in the reactor called *Arundo2*.

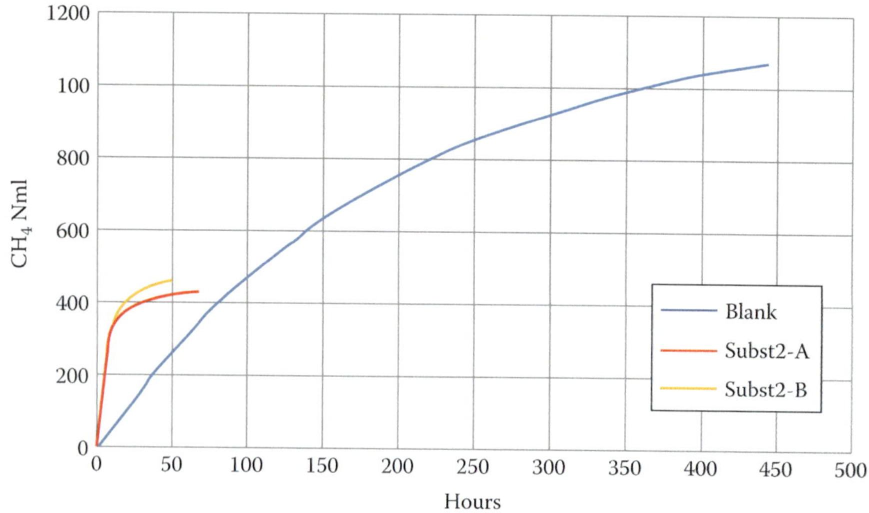

FIGURE 3.7
Example of a test failed because of method errors.

measuring the BMP of some agro-industrial waste (presumably containing a certain percentage of proteins and fats). The student decided to carry out his tests with I/S = 2 because it is the most frequent test condition in the literature (although this is questionable). The inoculum employed was sampled from a biogas plant running on cow manure and corn silage. Figure 3.7 illustrates the test results.

It is evident that such a test is wrong, and in this case, it is not necessary to investigate the causes like a detective. We can discard instrumental problems because the test was performed with a brand-new AMPTS II, perfectly calibrated and checked before the start. Gas leaks can be excluded too because both curves are coincident and interrupted at the same time. By analyzing the method adopted by the student, one finds the following errors:

1 General test planning: if one suspects that a substrate may contain fats, it is advisable to perform the test with I/S ≥3, to avoid provoking the "indigestion" of the bacteria. The Italian UNI/TS 11703/2018 standard recommends I/S = 5 in such cases.

2 The inoculum had not been preincubated, as can be easily deduced from the large amount of gas produced by the blank and the shape of the blank's curve. A well-preincubated inoculum should yield a very flat and almost linear curve.

3 The test was carried out in duplicate, which in this case increases the uncertainty.

4 The inoculum was sampled from a biogas plant that was usually running on cow manure and corn silage, i.e., high carbohydrate content and low protein and fats. The substrate to test was presumably composed of the inverse mix: little carbohydrates and much protein and fats. Therefore, even in the supposition that factors (1) and (2) had been correctly planned, the inoculum lacked specificity anyway. A two-step test would have been necessary, employing the digestate of the first test as inoculum for the second, to allow the bacterial flora to adapt to the new feedstock.

5 The quantity of inoculum loaded in the blank (400 mL) was different from that loaded in the replicates (150 mL). Even if the test had not been interrupted by the inhibition of the inoculum, a direct calculation of the BMP would have been impossible, requiring the inclusion of a correction factor and hence amplifying the errors in the calculation.

When checking the procedure for determining the VS, it was found that the student used just 1 g of sample "because the laboratory has a precision scale". The DW turned out to be 0.250 g, and the ash 0.055 g. On checking the scale's manual, it was found that the instrument's sensitivity was 0.005 g, but its accuracy was 0.02%. Remember that, by international convention, the accuracy is always expressed as a percentage of the end of scale (110 g for this concrete model). Consequently, the absolute error of said scale is 0.02% of 110 g, i.e., ±0.022 g. Calculating the error of the VS with the formula already explained in Section 3.2.3 (repeated here for convenience):

$$e_{\mathrm{VS}} = \frac{2 \cdot E_{\mathrm{scale}}}{\mathrm{DW} - \mathrm{Ash}} + \frac{E_{\mathrm{scale}}}{\mathrm{WW}},$$

we obtain the following results:

$$e_{\mathrm{VS}} = \frac{2 \cdot 0.022}{0.250 - 0.055} + \frac{0.022}{1} = \pm 25\,\%$$

When determining the VS of the inoculum, the student did a better job, but the result was still unacceptable. The WW was 5 g, resulting in just 0.380 g DW, while the ash was 0.035 g. Hence

$$e_{\mathrm{VS2}} = \frac{2 \cdot 0.022}{0.380 - 0.035} + \frac{0.022}{5} = \pm 13\,\%$$

Therefore, the error in the determination of the I/S is

$$e_{\mathrm{I/S}} = 25\,\% + 13\,\% = \pm 38\,\%$$

7 Such an error in determining the VS is already unacceptable, and it is amplified when calculating the I/S ratio. In practice, we cannot be sure what the I/S was, since any value between 1.24 and 2.76 could be theoretically possible. We can reasonably conclude that the sequence of errors led to the test being performed with an effective I/S <2. Hence, the sudden interruption of the methane production after only 60 h resulted from the inhibition of the bacterial activity, caused simultaneously by the excess of substrate and by the scarce specificity of the bacterial flora to digest a very "heavy" substrate, composed mainly of protein and fats.

3.10 The Hydrolytic and Methanogenic Activity Tests: Checking the Bacteria's Health

The procedure described in Sections 3.8 and 3.9 applies to any substrate whose BMP is unknown. A particular case of the BMP test is measuring reference substrates, i.e., pure substances whose BMP is known in advance.

This test helps compare different instruments (assuming the inoculum is the same for both and adequate for the substrate). It is advisable as practice for students and beginners (it is easy to find out any eventual problem in case of inconsistent results). For the biogas plant manager, it is a crucial test to assess whether the different microbial communities are healthy and consistent in number. Recalling the explanations in Chapter 1, Figure 1.2, and Section 1.1.2.2, the degradation of organic matter and its conversion into biogas is a sequential process. To test each link of the chain, there is a reference substrate for an AD assay (polysaccharides, proteins, lipids, and VFA). If the amount of methane obtained during the said tests is lower than a reference value, or even null, it means that the population of each specific bacterial group is inhibited or even missing.

Table 3.3 provides a list of reference substrates and their ranges of acceptable BMP, along with the groups of bacteria that can metabolize each one.

In the following paragraph, we will analyze, case by case, the peculiarities of each bacterial activity test compared to the generic procedure of the BMP test.

3.10.1 Degradation of Polysaccharides (Glucose, Starch, and Cellulose)

As shown in Table 3.3, all these substances have a reference BMP in the range of 350–400 Ncm^3/g VS. The test is easy to perform because the said substrates are easily digestible. To save money, it is possible to employ industrial products available in any supermarket, rather than their "laboratory grade" equivalents: white sugar or glucose syrup, corn, potato, or rice starch, and

TABLE 3.3

Reference Substrates for Biological Activity Tests

Reference Substrate	BMP Range (from Literature)	Specific Group of Bacteria
Cellulose	352–410 Ncm^3/g VS	Hydrolytic bacteria
Starch	370–410 Ncm^3/g VS	Hydrolytic bacteria
Glucose	350–373 Ncm^3/g VS	Acidogenic bacteria
Casein or gelatine	370–470 Ncm^3/g VS	Proteolytic bacteria
Propionic acid	330–350 Ncm^3/g COD	Acetogenic bacteria
Butyric acid	330–350 Ncm^3/g COD	Acetogenic bacteria
Fatty acids (vegetable oil)	800–1100 Ncm^3/g anhydrous	Lipolytic bacteria
Acetic acid	330–350 Ncm^3/g COD	Acetoclastic archaea
Sodium acetate (anhydrous)	359–273 Ncm^3/g anhydrous (330–350 Ncm^3/g COD)	Acetoclastic archaea

paper tissue or cotton. The reference BMP measured with such commercial substances may result in being closer to 350 Ncm^3/g VS than to 400 Ncm^3/g SV, because they are not 100% pure. It is enough to account for the purity grade, if indicated on the label, as explained below:

- *Glucose*: Usually sold in supermarkets as a syrup, it contains some water. Consequently, its theoretical BMP will be 373 Ncm^3/g VS multiplied by the purity percentage declared on the label.
- *White sugar*: Mostly composed of sucrose (saccharose), which is a disaccharide (two glucose molecules bound together). It may contain some additives and moisture, but in general, it is almost pure, hence we can assume its theoretical BMP to be equal to that of pure glucose.
- *Fructose*: Same chemical composition as glucose, but it tends to absorb more moisture than white sugar; hence, it is necessary to check thoroughly its TS and VS before calculating the quantity for each reactor.
- *Starch*: It is a polysaccharide (several glucose molecules bound together), and its chemical composition is quite variable. For example, cornstarch available in supermarkets contains, according to the manufacturer, "87% carbohydrates", with the rest being primarily moisture. Therefore, tests performed with this substrate should yield less than 357 Ncm^3/g on WW. Rice starch, on the other hand, is usually available in the "baby care" or "diet products" sections of supermarkets, with a purity typically higher than 99%. Being a hygroscopic substance, it is advisable to measure its TS each time, before starting the AD test.
- *Cellulose*: It is a polysaccharide, and its generic chemical formula is $(C_6H_{10}O_5)_n$. Pure microcrystalline cellulose (in powder) is available at shops specialized in laboratory materials and reagents. Industrial cotton (either as flocks or as disks) is composed of 99% cellulose. To increase its degradability, it is necessary to chop its fibers finely; this operation is more effortless with cotton in flocks. Ideally, the fibers should be

shortened and broken up in a way that avoids the formation of clumps when adding them to the inoculum. Tissue paper is instead composed of short fibers (to facilitate its disintegration in the septic tanks and sewage ducts). It is easier to employ than cotton, but its cellulose percentage is lower, about 80%–90%, depending on the quality. Furthermore, tissue paper can contain traces of Cl or other additives potentially inhibitors for the anaerobic process. Three-layer "soft rolls" are generally composed of 90% pure cellulose, and their fibers are short enough for easy comminution, making this quality of tissue paper a good reference substrate for AD tests. Filter paper employed in routine laboratory filtration is pure cellulose and yields BPM values closer to the theoretical maximum.

3.10.2 Casein and Gelatine in Powder

Both products are composed of pure proteins.

The elementary analysis of proteins in general shows the following composition: 55% carbon, 7% hydrogen, and 16% nitrogen. Proteins are highly complex and large molecules, some of which have a molecular weight of up to 10,000. They all differ from each other in terms of spatial structure, although their average elementary composition is quite similar to the values indicated.

Casein is extracted from milk whey. It is available in laboratory supply shops and as a food supplement in some shops specializing in dietetic products for bodybuilders and athletes. Gelatine powder is available in almost all supermarkets. The one employed for the preparation of desserts usually contains sugar, colorants, and flavors, and hence it is not suitable for our scope. Neutral gelatine in powder (the one employed for the preparation of galantine, aspic, and similar dishes) is the best alternative as a reference substrate for AD tests, because it dissolves easily and can be dosed quite accurately in each reactor. The time required for anaerobic degradation of proteins is generally long (about 20 days), and not all inocula can fully digest them in 30 days ($BMP_{30} < 370\ Ncm^3/g\ VS$).

3.10.3 Propionic and Butyric Acids

Propionic acid ($C_3H_6O_2$) and butyric acid ($C_4H_8O_2$) are both liquids having a rather pungent smell (especially the second, which is the usual component of rancid butter and vomit). Both are available only in specialized shops of laboratory supplies. Because it is well known that acids consume the alkalinity of the inoculum, it is a good practice to neutralize them either with caustic soda (NaOH) or with sodium bicarbonate ($NaHCO_3$). Consequently, the bacteria will digest sodium propionate or butyrate ($C_3H_5O_2Na$ or $C_4H_7O_2Na$). Sodium propionate is a chemical commodity, known in the food industry with the (European) code E281. It is a widely used preservative for industrial bread due to its anti-mold properties, and is commercially available in specialized industrial food supply stores. Sodium butyrate is an ingredient employed in pharmaceutical laboratories for the preparation of some perfumes and homeopathic medicines.

The theoretical BMP of both propionic and butyric acids is the same as that of acetic acid, i.e., 350 Ncm^3/g COD. Sodium propionate yields 530 Ncm^3 CH_4/g VS, while sodium butyrate yields 515 Ncm^3 CH_4/g VS.

To calculate the I/S when employing either of the acids (and hence, when it is impossible to determine the VS), the COD concentration of pure acid can be calculated as follows:

$$C_3H_6O_2 = 112 \text{ g COD/74 g acid} = 1.51 \text{ g COD/g propionic acid}$$

$$C_4H_8O_2 = 160 \text{ g COD/88 g acid} = 1.81 \text{ g COD/g butyric acid}$$

Performing the test

1 Filter the inoculum with a mesh (advised 5 mm, in any case smaller than 10 mm) to separate fibers and coarse solids.
2 Prepare a mixture of 0.5 g of propionic acid and 0.5 g of butyric acid, per liter of inoculum; hence, in total 1.66 g of total COD.
3 Prepare a blank reactor and a sample reactor loaded with the previously mentioned mixture. Start the experiment and check that the net production of methane is in the range of 435–581 Ncm^3 /g anhydrous product (350 Ncm^3/g COD).

N.B.: In this test, the usual I/S criterion (I/S = 2 or 3) is not valid, because the presence of propionic and butyric acid in proportions bigger than 1 g/L can inhibit the methanogenic archaea. Furthermore, the inoculum could already contain a certain amount of both acids (especially if it is a "fresh", non-degassed inoculum). As a reference guide, the mixture of acids described here has 1.66 g of COD, while the filtered inoculum may contain between 30 and 40 g COD/L; hence, the I/S ratio for this concrete test is greater than 18.

3.10.4 Fatty Acids (Lipids)

Lipids (oils and fats, either animal or vegetal) are very heavy molecules with more than 6 C atoms. Their high BMP has easily deceived many incautious biogas plant managers, with rather disastrous consequences—a biologically collapsed biogas plant or, at the very least, a significant loss of productivity. It is essential to note that such high BMP values, often cited in the scientific literature, are sometimes unattainable in industrial biogas plants, especially when their HRT is less than 45 days. Furthermore, since lipids degrade slowly and are hydrophobic substances, they tend to form emulsions, or worse, to float, preventing the inoculum's bacteria from digesting them. Lipids may inhibit some inocula (those derived from cow manure), while others (sewage sludge, swine manure) may easily digest them. For instance, stearic and oleic acids, which are typical components of vegetable oils with 18 carbon atoms, are reported to be toxic to the AD process at concentrations as low as 1 g/L. A dose in the range of 80–130 mg/L may already produce some inhibition in plants typically fed with other types of substrates.

The equivalence coefficients between VS and COD introduced in the former chapter are not valid for lipids. The reason is that lipids are composed mainly of carbon and hydrogen, so their total oxidation requires more oxygen than any other feedstock. For instance, consider oleic acid (the main component of olive and sunflower oils). Its total oxidation will undergo the following reaction:

$$C_{18}H_{34}O_2 + 25.5\ O_2 \rightarrow 18\ CO_2 + 17\ H_2O$$
$$282g + 816g \rightarrow 792g + 306g$$

Hence, the amount of COD per g of oleic acid is given by the following proportion:

$$\frac{816\left[g_{COD}\right]}{282\left[g_{oleic}\right]} = 2.89$$

To test the capacity to digest lipids, the rule of thumb is to consider the COD of oils and fats as three times their DW, and to choose the I/S in terms of COD, considering the inoculum's COD equal to 1.333 times its VS, expressed in g/L. With such a rule, the final concentration of oil and fats will be smaller than their usual inhibition limits, i.e., 0.6 g/L. Another rule of thumb is to consider the anhydrous weight of oil as total VS, and adopt I/S = 5 or higher. We must expect a long digestion time to observe an appreciable methane quantity in our gas flow meter. Furthermore, sometimes the shape of the curve will look more like a straight line than like a sigmoid. Inocula from sewage treatment plants are usually more suitable for digesting lipids than those from agricultural plants.

Example: We aim to test the lipolytic capacity of a given inoculum containing 5% VS. The reference substrate chosen is olive oil, and, for simplicity, its composition is assumed to be 100% oleic acid.

Therefore, the inoculum contains a quantity of VS equal to 5% of 1000 g/L, hence 50 g/L. The VS concentration, expressed as COD, is then 1.333 times the said value, i.e., 67 g/L.

Suppose we want to perform the test with 0.2 g of olive oil per liter of inoculum (toxicity limit of oleic acid according to Angelidaki and Ahring).

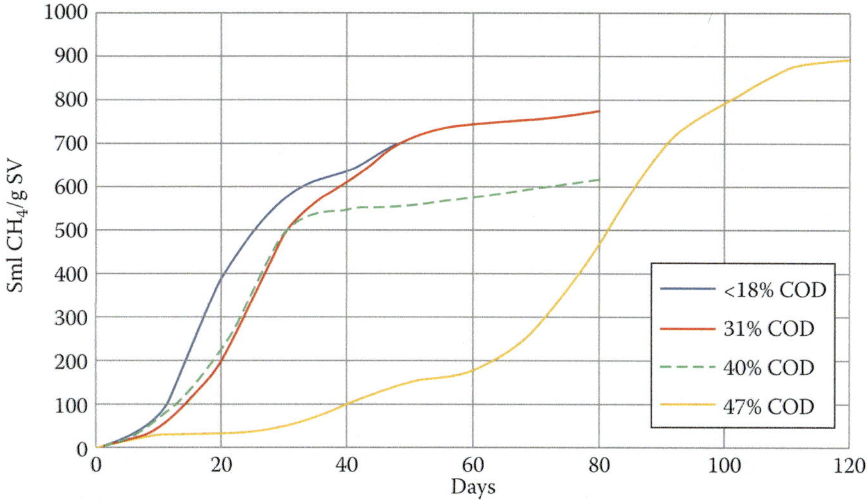

FIGURE 3.8
Degradation of triolein, according to Cirne et al. (2007). Graphics simplified by the author. According to the cited study, all tests where the amount of fat was less than 18% of total COD gave coincident curves. Tests with 31% of total COD showed some inhibition, but ultimately recovered and yielded substantially as much methane as the former. Tests with 40% COD showed inhibition and did not recover, yielding less methane than the previous tests. Tests with 45% of total COD yielded the highest BMP, but took 120 days.

Such a dosage is equivalent to 0.6 g of COD of oleic acid, 50 g COD/L inoculum in total, and hence I/S = 83.

Observe that, in general, oils and fats are not composed of pure oleic acid, but contain other fatty acids that are not so potent inhibitors of the AD process. For example, Cirne et al. found that the lower toxicity threshold of triolein (a triglyceride, comprising 75% of olive oil) becomes evident at an I/S ratio of 2.2 (in terms of COD). Consequently, in our hypothetical test with olive oil, we could safely add up to 5.5 g of oil/L of inoculum [hence I/S = 50/(5.5 × 3) = 3].

Since the exact composition of animal and vegetable fats is always uncertain, as well as the specificity of the inoculum to digest them, it is advisable to perform several tests with increasing I/S ratios, e.g., 3, 5, 10, and 15, and see which ones suffer inhibition. Remember that the duration of such tests must always be at least 45 days, and sometimes up to 60 days, as demonstrated in Figure 3.8.

3.10.5 Acetic Acid and Sodium Acetate

Both substrates, direct precursors of methane by the acetoclastic group of archaea, enable the carrying out of two complementary tests: the Specific Methanogenic Activity (SMA) test and the methanogenic capacity test. Both are useful for understanding whether the inoculum is suitable or not, or if the biogas plant has a sufficient resilience margin against inhibiting agents, or if the bacterial ecosystem is collapsing. The general philosophy of the SMA test assumes that, since acetoclastic archaea produce the majority of methane, adding acetic acid (or sodium acetate) to a given inoculum should result in consistent methane production. If not, it means that the inoculum contains little or no archaea, or that the same are inhibited. Such a condition may arise when employing *aerobic* sludge from a sewage treatment plant, or in the case of diagnosing problems in a digester that is already suffering from a biological collapse (a process already inhibited by any cause).

The test consists in loading the batch reactor with inoculum and a specified quantity of acetic acid, or sodium acetate, and checking if the BMP reaches a value close to the theoretical one. This first phase is called *methanogenic capacity test*, and it determines whether the inoculum is capable of converting acetate into methane or not. The second phase, i.e., the *SMA test*, consists in assessing the maximum methane flow generated by the inoculum in the presence of acetic acid or sodium acetate, as a function of the mass of (presumably) live and active bacteria (i.e., the VS of the inoculum).

From the same test, we will then be able to obtain two different insights: the methanogenic capacity and the SMA, respectively. The said information is necessary to assess whether the last link of the trophic chain in the AD process (methanogenesis) is as robust as the chain itself, or if it is the weakest step or bottleneck of the whole process. At this point, we need a short theoretical digression before studying some practical examples.

3.10.5.1 Stoichiometry of Acetic Acid and Sodium Acetate

The anaerobic degradation of acetic acid can be described with the following chemical formulas:

$$
\begin{array}{cccc}
CH_3COOH & \rightarrow \quad CH_4 & + & CO_2 \\
60\,\text{g acetic acid} & 16\,\text{g} = 22.4\ \text{Nl} & + & 44\ \text{g} = 22.4\ \text{Nl}
\end{array}
$$

Furthermore

$$60\,\text{g acetic acid} = 64\ \text{g COD}$$

then

$$BMP_{\text{theoretical}} = 22{,}400\ \text{Ncm}^3/64\,\text{g COD} = 350\ \text{Ncm}^3/\text{g COD}$$

The same reasoning applies also to sodium acetate

$$BMP_{\text{theoretical}} = 22{,}400\ \text{Ncm}^3/60\ \text{g VS} = 373\ \text{Ncm}^3/\text{g VS}$$

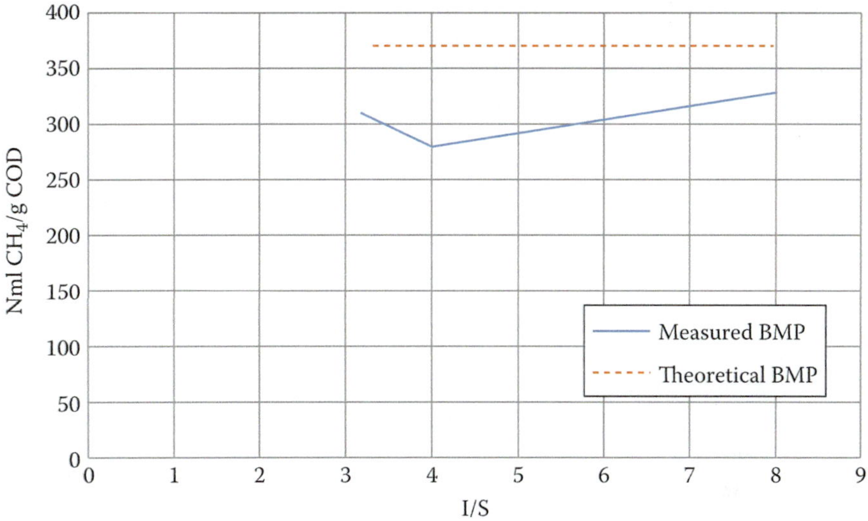

FIGURE 3.9
Variation of the BMP test results when employing sodium acetate, as a function of the I/S ratio. (Data from Badshah et al.; graphics by the author.)

3.10.5.2 First Step: Checking the Methanogenic Capacity of Acetate or Acetic Acid

The test follows the same general procedure as the BMP test. Special attention must be paid to the I/S ratio, because it has a direct influence on the measured value. Experimental evidence shows (Badshah et al.) that employing sodium acetate as a reference substrate with I/S < 3 will yield less methane than the theoretical production. The reason is the partial inhibition caused by the higher concentration of sodium. If employing acetic acid instead of sodium acetate, the results will be similar, in this case because of the partial inhibition caused by lowering the pH.

Figure 3.9 shows how the net methane production varies as a function of the I/S ratio when employing sodium acetate as the reference substrate, according to Badshah et al. (graphic elaboration by the author).

Observe that the experimental error will be minimum when 2 < I/S < 3, because the production of methane will be maximum, but at the same time, the concentration of sodium will be maximum too, and hence, the BMP could be smaller than the theoretical value because of the partial inhibition caused by the sodium. The test can last from 1 to 5 days.

By increasing the I/S ratio, the concentration of sodium drops, and its inhibitory effect drops accordingly. The duration of the test is usually shorter, 2 or 3 days, but the experimental error of the test increases, especially if the inoculum is too "fresh", i.e., insufficiently degassed.

As anticipated at the beginning of this paragraph, the test performed in the way described can be called "methanogenic capacity test", since it shows *if* the inoculum is capable of digesting acetate or not. It is not an SMA test, as it does not account for the time required for complete acetate degradation.

The following protocol outlines the procedure for conducting the methanogenic capacity test using sodium acetate.

- Sample the inoculum from the digester and sieve it through a 5-mm mesh to eliminate macroscopic fibers and solids.
- Measure the VS or the COD of the filtered inoculum (it is assumed that the measured VS represents the living bacterial biomass).
- Neutralize 6 g of acetic acid with 4 g of caustic soda diluted in 80 mL of distilled water. ATTENTION: This is an exothermic reaction. Wear gloves and eye protection.
- *Alternative*: Purchase anhydrous sodium acetate from a supplier of laboratory chemicals. Even if it is labeled as "anhydrous", it adsorbs moisture once the pot has been opened. To minimize errors in test preparation, bring the sodium acetate to the oven at 140°C and dehydrate it for 1 hour.
- Complete with distilled water up to 100 mL (or weigh 10 g of sodium acetate as soon as you take it from the oven and dilute it in 100 mL of distilled water). The resulting solution will then have 64,000 mg COD/L.
- Prepare a sample reactor with I/S = 5–8 and a blank reactor with the same quantity of degassed inoculum. Add distilled water until the total volume in both reactors is the same. Should the inoculum be "fresh", then preincubate it to minimize the error induced by the endogenous methane production.
- Check that the net methane production reaches 330–350 Ncm^3/g COD within 2 or 3 days (if I/S > 5) or 5 days (if I/S = 3). If it takes more time, it means that the inoculum has methanogenic activity, but it is inhibited.

It is possible to perform the former test with wine or apple vinegar instead of acetic acid. However, the uncertainty will be higher because the exact acidity depends on the product and is known only approximately. The following procedure allows you to prepare the test using vinegar as substrate:

Wine vinegar contains approximately 6% acetic acid; therefore, it is necessary to add 4 g of caustic soda per 100 g of vinegar to neutralize its pH. Start by adding and perfectly dissolving 3 g of caustic soda. Then, check that the resulting pH is less than 7 and add the remaining caustic soda in small quantities at a time, checking the pH until it reaches at least 6.8, ideally 7. The resulting theoretical COD will be 6400 mg/0.1 l (i.e., 64,000 mg/L) and conversely, the resulting theoretical VS will be 6 g/0.1 l, i.e., 6%. Should you need more or less caustic soda to reach the neutral pH, adjust the reference value proportionally (64,000 mg/L). For instance, if the pH reaches 7 with just 3.8 g of caustic soda, then the COD will be

$$COD_{real} = 3.8\ g/4\ g \cdot 64{,}000\ mg/L = 60{,}800\ mg/L$$

and the corresponding VS will be

$$VS_{real} = 3.8\ g/4\ g \cdot 6\% = 5.7\%$$

3.10.5.3 Second Step: The SMA Calculation

In this case, the scope is not measuring *how much* methane the inoculum can produce from acetate, but rather *at what speed* the bacteria degrade the substrate. Hence, instead of focusing on the *total production* of methane (Ncm^3/g COD or Ncm^3/g VS), we will measure the *maximum daily flow per g of VS of the inoculum*. This is the reason why the test is called "*specific* methanogenic activity".

The test protocol is analogous to the one already described in Section 3.10.5.2. Adopt an I/S > 5, typically I/S = 8 (if employing sodium acetate), otherwise employ acetic acid or vinegar, in such quantity as not to exceed 1 g acetic acid/L of mixture (when adding the vinegar or acetic acid directly to the inoculum). Check the daily net amount of methane produced. The said value must be divided by the *total VS of the inoculum*. In the case of agricultural digesters, the maximum quotient's value should be reached within 2 to 3 days. It must comprise between 30 and 40 $Ncm^3/g\ VS_{inoculum} \cdot d$, with an "acceptable" minimum of 10 $Ncm^3/g\ VS_{inoculum} \cdot d$. The latter value in $Ncm^3/g\ VS_{inoculum} \cdot d$ was deduced from Bolzonella (2013), who expresses the SMA in terms of COD. Other authors (Sandoval Lozano et al., 2009) report "usual" SMA values ranging from 17 to 70 $Ncm^3/g\ VS_{inoculum} \cdot d$ (these values are common in the case of anaerobic sewage sludge or industrial effluent treatment plants). In the case of granular sludge (typical of upflow anaerobic sludge blanket [UASB] or similar digesters), the SMA can reach values in the range 280–1,400 $Ncm^3/g\ VS_{inoculum} \cdot d$. In the extreme case of granular sludge, incubated at 60°C, the SMA reaches a peak at 2,100 $Ncm^3/g\ VS_{inoculum} \cdot d$ (Van Lier et al., 1995, already cited in Chapter 1, Figure 1.3). It is worth noting that in the wastewater treatment industry, many authors check the net methane flow on an hourly basis. On finding out the peak value, they then multiply it by 24 h, as a linear extrapolation. As an exercise, divide the indicative values of granular sludge quoted earlier by 24—the corrected range results comprise values between 12 and 58, which is very close to the range indicated by Sandoval Lozano et al. Whatever convention the reader will adopt, the important thing is to be coherent. Throughout this book, the author has adopted the first convention just because it was easier for him to deal with an Excel table reporting the daily flow rather than the hourly flow.

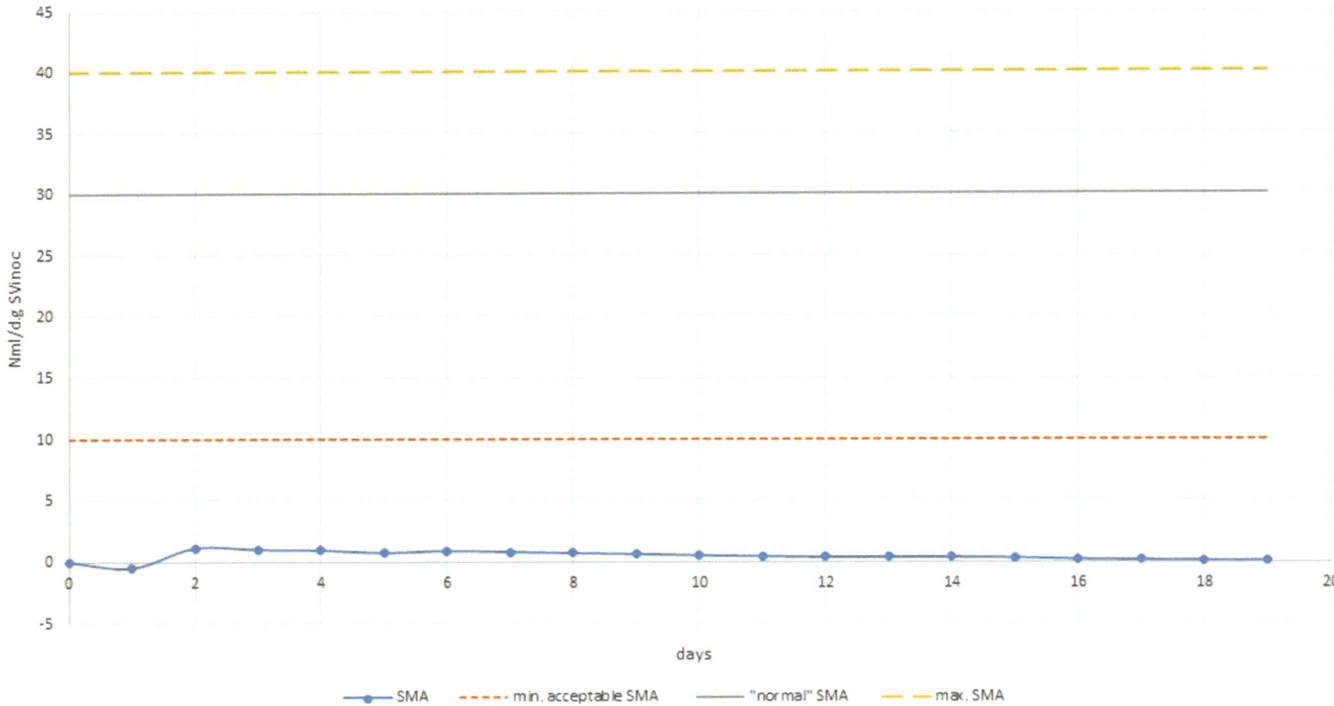

FIGURE 3.10
Example of inoculum with inadequate methanogenic activity, taken from a plant that was suffering biological problems. Results obtained with I/S = 5 and T = 38°C.

Important

In the scientific literature, it is frequent to express the SMA in terms of COD CH_4/g VS_{inoc}·d. The reason is that, to perform the balance of COD, it is easier to employ such a "scientific" unit. Nevertheless, laboratory instruments measure the volume of methane produced in a given time, hence Ncm^3/d. How to calculate the equivalence then? Remember that the quantity of methane produced by acetic acid or by sodium acetate (or by VFA in general) is 350 Ncm^3/g COD (Section 3.10.5.2); hence, 1 g COD of CH_4 = 350 Ncm^3 CH_4.

Figure 3.10 shows the results of an SMA test performed using an inoculum with inadequate methanogenic activity, as it was sampled from a digester experiencing some biological issues. Observe the negative value on the first day—this means that the blank produced more methane than the reactor with acetate, providing undeniable proof that the inoculum had already suffered from acidification, so adding more acetic acid was sufficient to inhibit it completely.

3.11 Analysis of the Error Propagation in the BMP Assay and Its Variants

At this point, we already know the measurement techniques of the parameters that are relevant for the biogas plant's management, and the theoretical BMP values of the reference substrates when the inoculum is "healthy". We must now calculate the uncertainty (aka error) percentage of our results to check whether we have performed our tests correctly.

By definition, the BMP of a given substrate is the result of the following formula:

$$\text{BMP} = \frac{\left(\bar{C} - \bar{B}\right)}{\text{VS}},$$

where

$\bar{C}$ = average of the normalized CH_4 production of the sample reactors

$\bar{B}$ = average of the normalized CH_4 production of the blank reactors (ideally almost null if the inoculum has been adequately degassed).

VS = volatile solids of the substrate (it is assumed that all the sample reactors have been loaded with the same amount of substrate VS).

N.B. The formula above is valid only if the quantity of inoculum in the blank and sample reactors is the same. It is also possible to use different amounts of inoculum in the sample and blank reactors by introducing a correction factor into the formula. Unless the substrate has a very low VS concentration (e.g., wastewater), we will always employ the same quantity of inoculum in the blank and sample reactors. If necessary, distilled water can be added to the inoculum in the blank reactors to have the same headspace volume in all reactors (a necessary condition for minimizing the error induced by the flush gas).

By applying the error propagation rules to the BMP formula, we get

$$e_{\text{BMP}} = \frac{e_{\text{instr}}\left(\bar{C} + \bar{B}\right)}{\left(\bar{C} - \bar{B}\right)} + e_{\text{VS}}$$

$$e_{\text{BMP}} \approx \frac{e_{\text{instr}}\left(\bar{C} + \bar{B}\right)}{\left(\bar{C} - \bar{B}\right)}$$

where

e_{instr} = error of the gas volume measurement instrument (including the normalization error if the instrument does not automatically normalize its reading)

e_{VS} = error in the determination of the VS, already explained in Section 3.2.3.

N.B. The numerator of the first term in the given formula is valid only if the single quantities of CH_4 produced by the sample reactors are approximately equal, and hence close to their average. In such a condition, it is possible to state that the absolute error of the average is equal to the relative error of the instrument, multiplied by the average itself. If the gas production of the single sample reactors differs significantly from the average value, the results should be discarded. The causes of said dispersion of values should be assessed, and then the test must be repeated under the correct conditions. Otherwise, the result obtained from averaging very dispersed single values would be unreliable, even if the calculated error may be acceptable.

Observe that, if the value $\bar{B}$ in the former formula was null (or anyway negligible compared to the value $\bar{C}$), then the first term will reduce to just e_{instr} (minimum possible error). This is the reason why it is mandatory (according to UNI/TS 11703:2018 and VDI4630:2016) to degas the inoculum with some days of preincubation before starting the test. If the biogas plant is running correctly, then the digestate will contain almost no more nutrients available for the bacteria. Hence, if the said digestate is sieved through a 5–10 mm mesh and employed as inoculum, its residual biogas production (aka "background noise" or "background production" of the test) will be minimal. Consequently, the preincubation could be omitted, on condition of accepting a slightly higher general error of the test.

To calculate the second term of the BMP error formula, it is then necessary to know the error of the instrument employed to measure the gas volume

TABLE 3.4

Typical Measurement Error Margins of Some Standard Instruments Diffused in the Market

Instrument	Max. Relative Error	Remarks
AMPTS II and III	±1% (+ e_{ga} if NaOH filters are not employed)	Normalization in real time, resolution 10 mL, can measure net methane or total biogas.
AMPTS II and III light	±1% (+ e_{ga} if NaOH filters are not employed)	Normalization in real time, resolution 10 mL, can measure net methane or total biogas.
Gas Endeavour II and III	±1% (+ e_{ga} if NaOH filters are not employed)	Normalization in real time, resolution 2 mL, can measure net methane or total biogas.
BPC Go	±1%+e_{ga}	Normalization in real time, resolution 10 or 2 mL, generic gas flow meter
MilliGascounter Cell	±(3%+e_b+e_t+e_{ga} if NaOH filters are not employed)	6% = min. normalization error assuming constant *T* and *P* for the normalization formula
Eudiometer and similar water displacement instruments	±(e_v+e_b+e_t)	e_v = error of the graduated cylinder or burette, calculated as the quotient between the cylinder's error class and the reading
Oxytop Bottle	±(1%+1hPa/P+e_T+e_v)	e_v=error in measuring the head volume of the bottle e_T=error of the temperature reading of the gas in the reactor's head space *P*=measured absolute pressure
Hohenheim syringes	±(5%-9%+e_b+e_t)	See an example of calculation in Chapter 6

e_b = error of the room barometer employed for the normalization of each reading
e_t = error of the room thermometer, idem
e_{ga} = error of the gas analyzer, including the uncertainty of the calibration mixture (when measuring total biogas production).

produced by the digestion. Table 3.4 shows the indicative errors of some instruments widely available on the market. Should the reader employ other instruments not listed here, he or she must take into account the maximum error specified by the manufacturer, plus the normalization error.

N.B. The errors of barometers and thermometers vary significantly from one model to another. Laboratory-grade electronic barometers usually have 2% error, while mercury column barometers typically have a 1.5%. The accuracy of glass capillary thermometers is typically ±1°C; therefore, when measuring mesophilic temperatures, the relative error is within the range of 2%–3%. Electronic laboratory thermometers typically have an absolute error of 0.1°C–0.2°C; therefore, their relative error when measuring mesophilic temperatures is in the range of 0.2%–0.3%. The error margin of gas analyzers is more challenging to assess, since it is a complex function of the error (tolerance) of the calibration mixture and the intrinsic measurement error of the instrument. The measurement error of the gas composition typically falls within the range of 2%–3%.

3.12 A Controversial Technique: Correcting the pH, the Alkalinity, and Adding Nutrients before Starting the Batch Tests

Correcting the pH and adding micronutrients (aka trace elements) to the inoculum is a widely adopted practice in the scientific world, often reported in the literature, and proposed by some authors as a standard procedure to follow at each test. The aqueous solution containing minerals, nutrients, and other substances that favor bacterial growth is usually referred to as a *medium*, *nutrient broth*, or *mother solution* in the scientific literature. From a scientific point of view, adding a solution that contains all the necessary buffers and nutrients for an optimal digestion process makes sense when the scope of the test is to determine the methane *potential* of a given biomass, i.e., the maximum amount of methane that the biological process (with a given inoculum) is *potentially* capable of extracting. If the goal is to measure the maximum value, it makes sense to put the bacteria in the best conditions to thrive. This means that the *medium* must contain all the mineral nutrients eventually lacking in both the substrate and the inoculum. In other words, the medium must provide enough buffer capacity to keep the pH as close as possible to 7, the organic nitrogen (biologically available, i.e., as nitrates and not in ammonia state), the micronutrients (P, K, Ca, Fe, Ni, and Co; according to some authors also traces of Se are necessary), vitamins, and other substances considered as probiotics for the microorganisms. In practice, the *medium's* formula should be decided in each specific case, depending on the substrate and inoculum employed for the test. The German norm VDI4630/2016 and the Italian norm UNI/TS11703:2018 state that no buffer and probiotics should be added to the inoculum unless it is known, or suspected, that the inoculum lacks such compounds. The Italian norm includes a recipe for a mother solution and the minimum amount to be added to the inoculum. The criterion adopted is to ensure that the microorganisms have a minimum amount of essential mineral nutrients, thereby avoiding underestimation of the BMP.

In the author's personal opinion, both norms are strongly biased by the personal background of the people who wrote them. The German norm is too focused on agricultural biogas plants. In Germany, such biogas plants are primarily fed with more cereals than manure, and it is common practice to regularly add micronutrients to the digester to maintain the health of its bacterial ecosystem. Hence, it is pretty logical that adding micronutrients to the inoculum when performing a laboratory test—as advocated by many authors—is not necessary when the inoculum has been collected from such digesters, because it already has the required concentration of micronutrients. In some cases, the addition of medium in the laboratory can be detrimental because the excess of certain minerals (Ni, Co, and Fe) can inhibit the hydrogenotrophic archaea, resulting in a lower methane yield. On the other hand, if the scope of the test is to obtain information for managing a biogas plant fed with sludge from sewage treatment, it is often the case that such inoculum may lack some essential mineral nutrients. It is then necessary to add medium to the inoculum before starting the laboratory test to avoid the underestimation of the BMP. Nevertheless, the BMP obtained in this way is biased because it does not accurately reflect the effective state of the inoculum within the digester. From a practical point of view, it is more useful always to include a control reactor loaded with acetic acid or sodium acetate. If the specific methanogenic activity (SMA) is greater than the acceptable minimum (10 $Ncm^3/g\ VS_{inoc} \cdot day$), then the plant's inoculum is "healthy" and the test is acceptable; otherwise, it must be repeated with the addition of micronutrients and the digester needs to be "boosted" as well.

It is always advisable to perform a comparative test from time to time, to check if the addition of medium to the inoculum results in a higher methane yield from the feedstock (especially if this latter contains anti-nutrient factors), or if, on the contrary, it turns out to be inhibitory. It is imperative to check the error propagation of the test. Quite often, the author has read scientific papers where "improvements" in methane yield of 3% or more are presented as "positive results", demonstrating the efficacy of micronutrient addition. We know from experience that it is challenging to perform a BMP test with less than 3% error when using highly sophisticated volumetric instruments with real-time normalization, with a minimum error of 1% currently possible with available technology. On the contrary, a 5% or more error margin is quite frequent when employing other instruments for the tests. When performing a comparative test, any difference of the same order of magnitude of the error margin is irrelevant, as it is impossible to state whether the real cause of the difference is the addition of the nutrients or the instrumental errors.

Figure 3.11 shows an example from real life. Two reactors were filled with 1800 mL each of fresh inoculum, sampled from the first reactor of a three-step agricultural biogas plant. One reactor was left unaltered, while the second reactor was added with medium according to the Italian norm. The difference between the two was around 8% less methane in the case of the reactor that received additional micronutrients.

The former test was performed with an AMPTS Light, having $e < 1\%$. The difference encountered is larger than the error margin of each single methane production. Hence, we can be certain that in this case, the addition of medium to the reactors led to partial inhibition due to excess micronutrient dosage. On the other hand, we cannot state with absolute certainty that the magnitude of

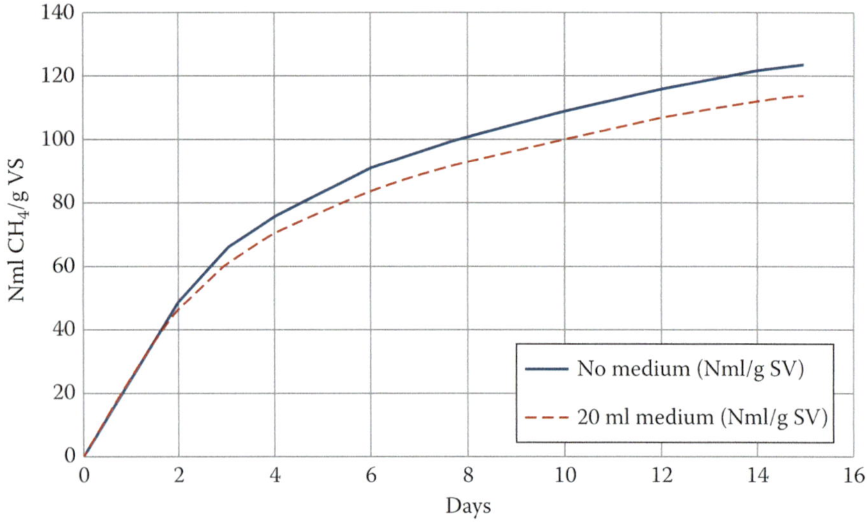

FIGURE 3.11
Negative influence of the excess of micronutrients in the inoculum.

the inhibition is 8%, because the difference between the two measurements, each one having 1% error, has its own uncertainty:

$$E_{(S1-S2)} = E_{S1} + E_{S2} = 1\%\left(123 + 113\right)\text{Ncm}^3/\text{g VS} = 2.36\ \text{Ncm}^3/\text{g VS}$$

Since the values of both measurements are very close to each other, the relative error will be high:

$$e_{(S1-S2)} = E_{(S1-S2)} / \left(S1 - S2\right) = 2.36/10 = 24\%$$

In this case, 24% is the uncertainty of the calculated difference.

In other words, we cannot be certain if the actual difference between adding nutrients to the inoculum or leaving it as is will be exactly 8%, since any value within the range of –6% to –10% could be possible. In any case, we can state for sure that the addition of medium to one of the reactors led to its partial inhibition, –8% being the "most probable" magnitude of said inhibition.

Bibliography

Angelidaki, I., and Ahring, B., Effects of free long-chain fatty acids on thermophilic anaerobic digestion, *Applied Microbiology Biotechnology*, *37*, 808–812, 1992.

Angelidaki, I., Alves, M., Bolzonella, D., Borzacconi, L., Campos, J., Guwy, A., Kalyuzhnyi, S., Jenicek, P., and van Lier, J.B., Defining the biomethane potential (BMP) of solid organic wastes and energy crops: A proposed protocol for batch assays, *Water Science & Technology*, *59*(5), 927–934, 2009.

Badshah, M., Liu, J., and Mathiasson, B., *The Biomethane Potential of Chemically Defined Substrates Evaluated Using Automatic Methane Potential Test System*. Lund University Publications, Lund, Sweden, 2012.

Bolzonella, D., *Monitoring and Lab Analysis—Assays*. Lectures on biogas technology, Jyväskylä Summer School, Jyväskylä, Sweden, 2013.

Cirne, D.G., Paloumeta, X., Bjornsson, L., Alves, M.M., and Mattiasson, B., Anaerobic digestion of lipid-rich waste—Effects of lipid concentration, *Renewable Energy*, *32*, 965–975, 2007.

Esposito, G., Frunzo, L., Liotta, F., Panico, A., and Pirozzi, F., Bio-methane potential tests to measure the biogas production from the digestion and co-digestion of complex organic substrates, *The Open Environmental Engineering Journal*, *5*, 1–8, 2012.

Farina, R., Bernard, O., Steyer, J.P., Conte, T., Lema, J., Roca, E., Ruiz, G., Franco, A., Cellamare, C., Spagni, A., and Martinez, J.A., Misassumptions in COD measurement in wastewaters containing high concentration of VFA, ENEA publication, *Protezione Unità Tecnica Valutazioni Ambientali—Laboratorio Protezione e Gestione della Risorsa Idrica*, 2004.

Ficara, E., Scaglione, D., and Caffaz, S., *Test di laboratorio per valutare la producibilità di biogas, Chapter 9*. In Vismara Renato, Malpei Francesca, Canziani Roberto & Piccinini Sergio (eds.). *Biogas da agrozootecnica e agroindustria*, Dario Flaccovio Editore, Palermo, Italy, 2011.

Porter, M.G., and Murray, R.S., The volatility of components of grass silage on oven drying and interrelationship between dry matter content estimated with different methods, *Grass Forage Science*, *56*, 405–411, 2001.

Sandoval Lozano, C. et al., Microbiological characterization and specific methanogenic activity of anaerobic sludge used in urban solid waste treatment, *Waste Management*, *29*, 704–711, 2009.

Schneiders, D., da Silva, J., Till, A., Lapa, K., and Pinheiro, A., Atividade methanegênica específica (AME) de lodos industriais provenientes do tratamento biológico aeróbio e anaeróbio, *Revista Ambiente & Água—An Interdisciplinary Journal of Applied Science*, *8*(2), 135, 2013.

Sung, T., and Alastair, D., Long chain fatty acids degradation in anaerobic digester: Thermodynamic equilibrium consideration, *Process Biochemistry*, *45*, 335–345, 2010.

VDI (Verein Deutscher Ingenieure), *Norm VDI 4630—Fermentation of organic materials. Characterisation of the substrate, sampling, collection of material data, fermentation tests*. Beuth Verlag GmbH, Berlin, 2016. www.dinmedia.de/de/technische-regel/vdi-4630/244849582.

4

Application of Laboratory Experimental Results to Biogas Plant Management

4.1 Practical Applications of the VS Test

The total amount of carbon, assumed *a priori* as "digestible", is exclusively contained in the solid fraction defined as volatile solids (VS). The VS test is straightforward to carry out and provides insightful information for the supervision and control of the anaerobic digestion (AD) process.

4.1.1 Measuring the Organic Load (OL) and the Organic Load Rate (OLR)

The first practical application of the VS measurement is to assess whether the plant's OL and the OLR are within their design limits.

Most of the agricultural biogas plants built in Europe belong to the bioreactor category known as "continuous stirred tank reactor" (CSTR). This type of digester cannot accept an OLR >3 kg of VS/m^3·day; otherwise, the biomass conversion efficiency into methane will drop to unacceptable levels. Typically, single-stage and two-stage biogas plants require an OLR in the range of 2.5–3 kg of VS/m^3·day. The so-called "compact digesters", promoted by some German plant builders, can work with up to 5 kg VS/m^3·day. According to the same manufacturers, the stability of the bacterial ecosystem requires a daily dose of "integrators" (mixtures of enzymes, minerals, and lyophilized bacteria) to maintain digestion efficiency at acceptable levels. The vendor's argument in favor of such digesters is that they do not belong to the CSTR type, but their working principle is that of the so-called plug flow digester (PFD), which is (theoretically) more efficient than the CSTR. Despite the theoretical considerations and the peer-reviewed literature, some of the author's customers in Italy had very negative experiences with such digesters because of the following reasons:

1 It is not true that the commercial compact digesters are "real" PFRs. In practice, their operation falls somewhat between that of an ideal PFR and a CSTR.

DOI: 10.1201/9781003676201-4

2 The cost of the "integrators" is high, since they are proprietary formulas of the same digester builder. Nevertheless, during the plant's commercial operation, the owners observed that the additives were not as effective in keeping an acceptable digestion efficiency as the manufacturer claimed. Such low efficacy was also demonstrated by batch tests in the author's laboratory: the residual BMP of the digestate collected at the exit of the PFR was approximately 50% of that of the feedstock.

In the case of conventional single-stage CSTR digesters fed with agricultural biomass, exceeding 3 kg of VS/m^3·day can result in acidification and eventual biological collapse of the AD process. In the case of biogas plants with two or more digesters, the OLR must be calculated by dividing the daily OL by the sum of the volumes of all the digesters.

4.1.1.1 Practical Application

It is advisable to perform the following tests at least once monthly:

1 Take a representative sample of the feedstock (or a representative sample of the mixture, if the digester's diet consists of a mixture of different biomasses) and measure its VS. In the case of corn or other silage, a representative sample can be prepared by taking a handful of silage from at least six different points along the trench's loading front, thoroughly mixing it in a container, and then taking the necessary amount of sample from the equalized mixture. In the case of sludge or liquid manure, it is advisable to thoroughly stir the collection pit with a stick before taking a sample from it. Always avoid sampling slurry from the surface or the bottom of the tank. Once you have a representative sample, measure its VS according to the ordinary procedure.

2 Calculate the OL with the usual formula:

$$\mathrm{OL} = Q_{\mathrm{biomass}}\left[\mathrm{ton/day}\right] \cdot \mathrm{VS}\left[\%\ \mathrm{of\ WW}\right]$$

3 Calculate the OLR with the usual formula:

$$\mathrm{OLR} = \frac{\mathrm{OL}}{\Sigma V_{\mathrm{digesters}}}.$$

Please note that if your plant has an airtight storage tank for storing the digestate, its volume must be included in the formula above, since such a tank can be considered as a post-digester.

4 Take a sample of digestate from the storage tank (if it is of the airtight type) or from the last digester (if your plant has an open digestate storage tank or basin). Measure the VS of the digestate.

5 Copy the measured values in a table or a spreadsheet [date of sampling, dry matter (DM), VS of the feedstock or biomass mix (VS_{in}), VS of the digestate (VS_{out}), OL, and OLR].

6 Plot the data as historical graphs (VS_{in}, VS_{out}, OL, and OLR on the y axis; dates on the x axis). This will clearly show the plant's trends. Please note that the silage's VS tends to decrease over time due to oxidation in the silo or trench, so it will be necessary to adjust the total amount of feedstock, Q, to maintain constant OL and energy production.

7 Should you notice that the VS_{in} decreases (for instance, because the silage absorbed moisture), then recalculate the quantity of feedstock, Q, to maintain the OL constant.

4.1.2 Measuring the Efficiency of the Feedstock's Conversion into Methane

The second practical application of the VS measurement is to assess the efficiency of biomass conversion into methane. This verification helps optimize the purchase cost of the feedstock, if applicable. Another application is to control the regular operation of the biogas plant, as managers often increase the OLR to stabilize biogas production. Still, the consequence of such action is reducing the net methane yield compared to the feedstock's BMP. *Increasing the OLR is like giving gas to a car's motor—the output power will increase, but the fuel consumption will increase more than proportionally.*

In biogas plants having an airtight digestate storage tank, the problem described earlier becomes marginal, because such tanks, being unheated, serve as psychrophilic post-digesters featuring long HRT. On the contrary, in biogas plants having an open storage tank, the incomplete digestion of the biomass caused by an increase in the OLR not only wastes feedstock but also emits more greenhouse gases into the atmosphere. Consequently, biogas plants lacking a gastight digestate storage tank require more frequent and accurate measurements of digestion efficiency. Figure 4.1, taken from a study performed in the United States in 1998, shows how the digestion efficiency decreases with increasing OLR (an inverse proportional relationship).

If the procedure described in Section 4.1.1 has been carried out correctly, the biogas plant's manager will have all the necessary elements to calculate the digestion efficiency. The latter is defined as:

$$\eta_{AD} = \frac{SV_{in} - SV_{out}}{SV_{in}}$$

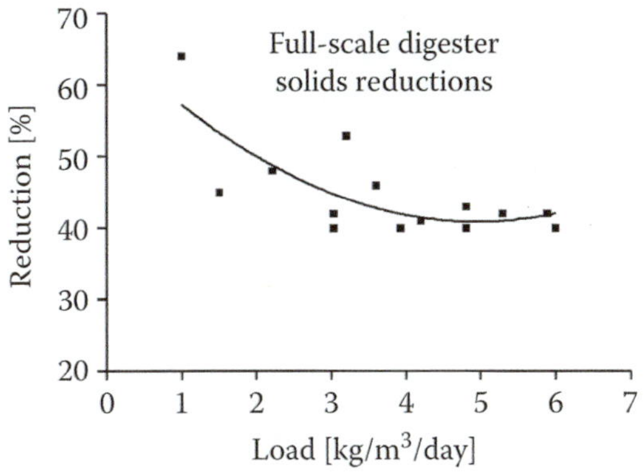

FIGURE 4.1
The efficiency of the feedstock's conversion into methane decreases with increasing the OLR of the biogas plant. (Burke, *Dairy Waste Anaerobic Digestion Handbook*, Environmental Energy Company, Olympia, WA, 2001, who in turn quotes Lusk, 1998. With permission.)

The procedure to monitor the plant based on the calculated η_{AD} is as follows:

1 Calculate η_{AD} with the said formula, using the data measured as explained in Section 4.1.1.
2 Save the information (date, calculated value of η_{AD}) in a file containing the historical series.
3 If $\eta_{AD} > 0.45$, then the operation of the plant is acceptable. Lower η_{AD} values could indicate the existence of a hydraulic short circuit (if the trend is a constant reduction of the efficiency) or temporary partial inhibition of the process (in the case of a single value being lower than the average). Feedstock containing a high percentage of lignin can also lead to low values of η_{AD}.

Table 4.1. Example of a spreadsheet, containing the necessary data to monitor the OL and the efficiency of the AD process. It is convenient to plot the values of VS_{in}, VS_{out}, OLR, and η_{DA} as a function of time in a separate sheet to detect negative trends easily.

4.1.3 Corrective Actions in the Case $\eta_{DA} < 45\%$

It is essential to note that η_{DA} seldom exceeds 60% in single-stage industrial biogas plants, as the microorganisms utilize part of the carbon in the feedstock to produce their own biomass (proteins and lipids accumulated within the cells) and to reproduce themselves. In CSTR-type plants, it can happen that, after 4 or 5 years of operation, the efficiency η_{DA} tends to gradually, but constantly, decrease despite the OLR being kept within design limits. There can be two causes: either the partial inhibition of one of the steps of the AD process (usually the methanogenesis) or the reduction of the hydraulic retention time (HRT). It is possible to diagnose the first case easily through the biological tests explained in Section 4.2.4. Should no biological problem arise from the said tests, then the most probable cause of the decrease in efficiency is a *hydraulic short circuit*, i.e., a reduction of the effective volume of the digester, caused by the accumulation of sediments in its bottom (as shown in Figure 4.2) and/or the formation of a mat of floating fibers on its surface.

Remember that, by definition:

$$\mathrm{SRT} = \frac{V_{\mathrm{digester}}}{Q_{\mathrm{in}}}$$

Therefore, if the input flow is kept constant to produce a given amount of energy, but the useful volume has become smaller because of the accumulated sediments, then the solids retention time (SRT) diminishes proportionally. Shorter SRT implies that the feedstock will be digested only partially. By

TABLE 4.1

Possible Template that the Reader Can Easily Prepare with a Spreadsheet

Week	DM_{in} [%]	DM_{out} [%]	VS_{in} [%]	VS_{out} [%]	OL	OLR	η_{DA} [%]	η_{DA} >0.45?
n	15	10	11.8	6	30 ton·VS/day	2.8 kg·VS/[m³·day]	49	OK
n + 1	13	11	11.5	6.4	30.1 ton·VS/day	2.8 kg·VS/[m³·day]	44	Check!

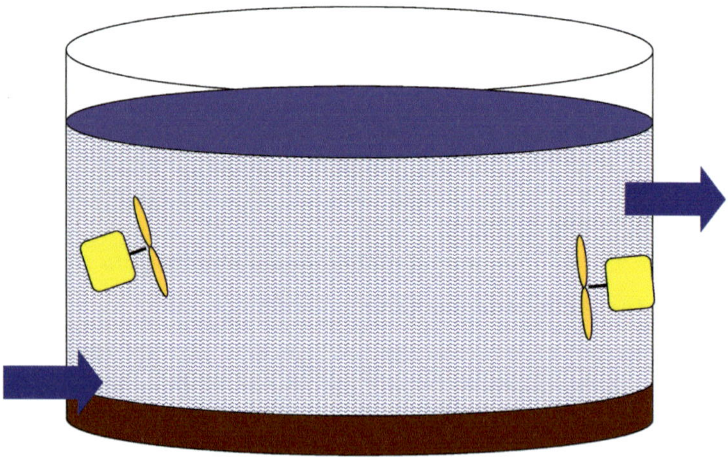

FIGURE 4.2
Reduction of the effective volume of the digester caused by sediment accumulation at the bottom.

periodically monitoring the VS and the digestion efficiency, $\eta_{DA,}$ one can pinpoint such issues well before they become critical.

Should the problem persist, the possible solutions are two:

1 Bring the digester's stirrers to their maximum power and closely monitor during the process. If the increased stirring can lift the sediments, then the digestion efficiency, η_{DA}, will begin to rise, until it reaches an acceptable limit.

2 If the former solution proves ineffective, it will be necessary to perform a test with chemical tracers, to check *if* the volume of the digester has been reduced and by *how much*. The test is relatively expensive and requires the assistance of an external chemical laboratory, both for determining the amount of tracer to load into the digester (at least 30 times the detection threshold of the available analytical instruments) and for analyzing 30 samples. The general idea consists of loading the digester with a predefined quantity of an inert chemical product, which is not contained in the feedstock. Typically, lithium bromide (LiBr or other salts of F, Br, and Li are employed.

If the digester's stirring was perfect (and we assume this because the stirrers would have been running at full power during the process), then the concentration of LiBr will decrease in time according to an exponential law of the form:

$$C = C_0 \left[\frac{\left(V / q^2\right)}{T} \right] e^{-(t/T)}$$

where

C_0 = initial concentration

V = effective volume of the digester [m^3]

q = feedstock's volumetric flow [m^3/day]

T = nominal (or theoretical, according to the case) HRT (Remember: HRT = SRT because this test is valid only for CSTR digesters)

t = day when the sample was taken (usually one sample a day, always at the same time, during the test's duration)

When monitoring daily the concentration C, the day when $C \approx 0$ (threshold limit of the analytical method employed) represents the *real* HRT of the digester. If a hydraulic short circuit effectively exists, then the measured HRT will be shorter than the nominal HRT of the plant.

Now we know the actual HRT, and since the daily feedstock's flow (assumed constant during the test duration) is known, then we can calculate the effective volume available in the digester (by just applying the definition of HRT). Having a clearer view of the problem's magnitude, the manager of the biogas plant can now decide with more certainty which of the following strategies to adopt:

1 If the biogas plant is still "young" (i.e., its commercial end of life is expected in many years in the future), then it may prove more convenient to stop operation, empty the digester, clean it, perform any other maintenance, and finally restart the plant.

2 Keep the stirring at maximum power for some months and check if the situation improves. It is a good practice to eliminate fibrous and high-solid materials from the digester's diet, to prevent the accumulation of indigestible matter, and to employ liquid biomasses (whey, sludge, glycerol, etc.) or easily degradable feedstock (oilseed cake, light food rests, nonedible flour).

A simpler and more cost-effective alternative to the test with tracers consists of taking out a vertical strip of thermal insulation from the digester's wall and performing thermography. Sediments or floating mats will appear as cooler regions in the thermography, allowing for the calculation of the exact residual digestion volume.

Whatever the method employed to estimate the lost digestion volume, taking the correct decision is not straightforward, requiring a thorough cost/benefit analysis on a case-by-case basis.

Practical Conclusions

By checking digestion efficiency at least monthly and maintaining a historical record, the biogas plant manager will be able to diagnose any issues as soon as they begin to manifest. In the case that a slight but steady negative trend of the efficiency values becomes measurable, then the problem can be easily solved by increasing the stirring power and/or changing to a less "fibrous" diet. When hydraulic short circuits are diagnosed promptly, there is a high probability of correcting the problem and preventing the accumulation of sediments and floating mats, whose removal may ultimately require shutting down the plant and opening the digester to clean it.

4.2 Practical Applications of the BMP Test

Regardless of whether measuring unknown biomasses or reference substrates whose BMP is known a priori, the procedure for carrying out this test remains the same, with some minor variations already explained in Chapter 3. The following practical examples will help one to understand the economic importance of the BMP assay.

4.2.1 The Right Price for the Feedstock

In many industrialized countries, investors build biogas plants primarily to produce energy, drawn by local policies that subsidize such installations. If this is the reader's situation, the following example will demonstrate how to optimize the feedstock's procurement activities.

A vendor offered one of the author's customers a "special feedstock with high biogas yield" at 150 €/ton. The customer's question was naturally: Is the energy production worth the purchase cost of the feedstock?

The "special formula" in question, referred to here as "Substance X", was a fine, dry powder with an almost imperceptible smell. Its characterization gave the following values:

- DM = 90.94 % on w.w.
- VS = 85.04 % on w.w.
- Ash = 5.9 % on w.w.

The inoculum employed for the test was digestate sampled from the same plant, previously filtered to eliminate coarse fibers. Its VS concentration turned out to be 3.8%.

The assay was performed using an AMPTS-II, which features real-time normalization of the measured methane volume, with its overestimation correction function activated. The tests were carried out in duplicate. The test temperature was set at 38.5°C (because the plant works in the range 38°C–39°C). The I/S ratio was chosen to be 3, resulting in 4 g VS of Substance X to be loaded in each reactor. Table 4.2 presents the individual methane volumes produced by each reactor, along with the average result (fifth column, which shows the calculation of specific production).

The net average methane yield is easier to analyze when plotted as a graph rather than in a table. Then the graphic function of the spreadsheet turns very useful, giving the result shown in Figure 4.3.

In general, the anaerobic degradability of such a substrate is good, as it yields 94% of its BMP within 12 days. The curve's shape is somehow anomalous if compared to the classical sigmoid. The causes could be either that the inoculum is not "specific enough" to digest such substrate, or that Substance X is a mixture containing easily degradable sugars, complex polysaccharides, and or/proteins (more difficult to digest), and probably a small proportion of fats. The test's duration was just 23 days because "the result was required urgently". Such a short duration raises doubts about the *ultimate* BMP. According to the vendor, the measured BMP was lower than his own estimation. Not without reticence, the vendor finally admitted that Substance X contains "some fats". Fats have a high BMP. Certainly, a small

TABLE 4.2

Results of the Individual Reactors and their Averaged Specific Production

Day	Blank [Ncm³]	X1 [Ncm³]	X2 [Ncm³]	Net CH_4 Production [Ncm³/g SV]
0	0	0	0	0
1	554	934	1061	111
2	725	1161	1286	125
3	1028	1499	1620	133
4	1368	1987	2131	173
5	1600	2366	2580	218
6	1716	2628	2868	258
7	1812	2812	3063	281
8	1932	2959	3204	287
9	2079	3116	3347	288
10	2126	3285	3510	318
11	2150	3328	3588	327
12	2174	3359	3616	328
13	2197	3387	3642	329
14	2219	3414	3667	330
15	2240	3442	3691	332
16	2260	3471	3717	333
17	2277	3498	3738	335
18	2293	3522	3763	337
19	2309	3545	3784	339
20	2325	3572	3808	341
21	2338	3594	3828	343
22	2346	3608	3846	345
23	2346	3608	3846	345

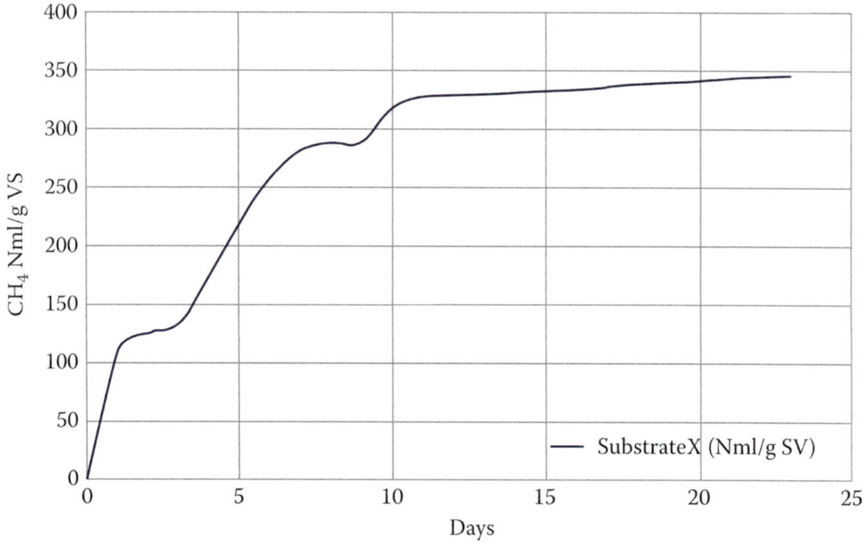

FIGURE 4.3
The complete anaerobic degradation curve of Substance X.

percentage of them would sensibly increase the BMP of the mixture, but on the other hand, their complete degradation requires 45–60 days. Then, at least 60 days of testing would have been necessary to determine the "ultimate" BMP of Substance X. On the other hand, the biogas plant is designed for an SRT = 30 days. So, the BMP value that really matters for the customer is the yield in 30 days, and the test with 23 days is a good approximation, given the "hurry" to decide whether to purchase the feedstock or not.

Now that we have all the data about Substance X's potential, calculating its economic convenience is relatively straightforward. Since the concentration of VS is 85%, then each ton of Substance X contains 850 kg, and will yield

$$Q_{CH_4} = 850\,[\text{kg/t w.w.}] \times 0.350\,[\text{Nm}^3/\text{kg}] = 297.5\,[\text{Nm}^3/\text{t w.w.}]$$

The direct cost of the methane produced with such feedstock will be:

$$C_4 = 150\ €/297.5\ \text{Nm}^3 = 0.504\ €/\text{Nm}^3$$

Considering that the generator has 38% electric efficiency,[1] and that the lower calorific value of methane is 9.94 kWh/m^3, the direct cost of the electricity generation will be:

$$C_4 = 0.504\ €/\text{m}^3/(0.38 \cdot 9.94\ \text{kWh/m}^3) = 0.134\ €/\text{kWh}$$

The said cost is then the "fuel cost" of the plant, supposing to feed it with Substance X. Adding the costs of mortgage, labor, lubricants, maintenance, etc., will result in the total cost per kWh, which must be then compared to the feed-in tariff, to decide whether the substrate in question is economically convenient or not.

In this special case, the vendor declared a BMP that makes the substrate look slightly profitable with the customer's feed-in tariff. The effective BMP measured in the lab, assuming 30 days or shorter HRT (real operational condition of the plant) was too low to justify the price required by the vendor.

4.2.2 Optimizing the Biogas Plant's Diet

Like any other living being, microorganisms will grow healthier if they are fed with an assorted and balanced diet. Furthermore, a rich and biodiverse bacterial ecosystem will improve the digestion efficiency of the plant. The following practical rules will help the biogas plant manager to optimize the digester's diet:

1. Mixtures of different substrates will generally work better than "pure" substrates (i.e., only carbohydrates, proteins, or fats).

2 2 + 2 will not always yield 4 in the AD reign. A mixture of two substrates will yield its maximum methane potential if their respective proportions are such that the C/N/P ratio results as near as possible to 150/5/1.

3 Mixtures of manure from different animal species, having different bacterial floras, will be more biodiverse than the pure manure of a single species. In general, such mixtures will be able to degrade a broader range of feedstocks.

4 Ruminants' manure (cows, buffalos, goats, sheep, camels, reindeer, etc.) is a suitable inoculum for digesting cellulose-rich substrates. Monogastric animals' manure (pigs, chickens, ducks, humans, etc.) is a suitable inoculum for digesting protein and fat-rich substrates. Manure from nonruminant herbivores (rabbits, horses, elephants, etc.) usually has a higher BMP than that of ruminants, because such animals have a different bacterial flora allowing them to digest cellulose and hemicellulose (Zhao et al., 2012), but are less efficient than cows in chewing their fodder, since they only chew once. Hence, the said manure is a good substrate and an acceptable inoculum to start a biogas plant too.

Figure 4.4 shows an example of how to empirically optimize the digester's diet. A biogas plant running regularly on cow manure had the opportunity to incorporate rabbit manure. The questions of the biogas plant's manager were

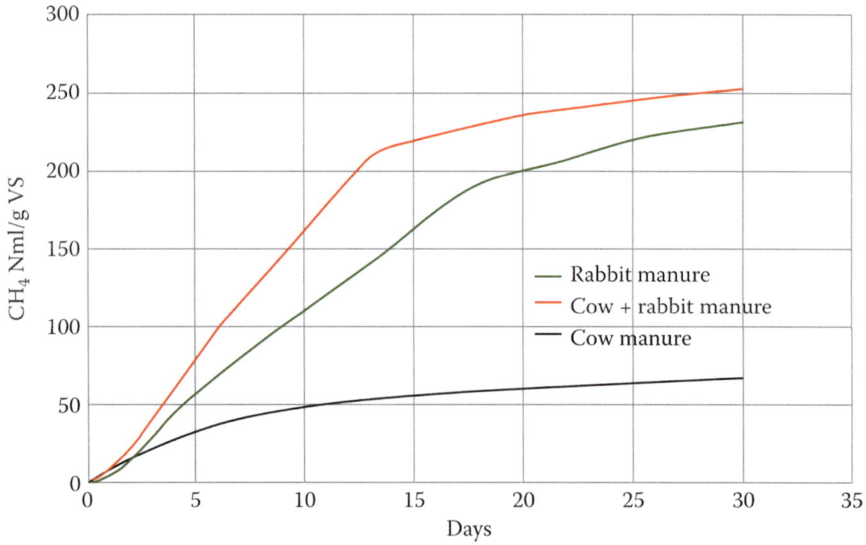

FIGURE 4.4
Example of the results that can be obtained by just mixing substrates in the right proportions.

"Is the addition of rabbit manure beneficial?"; "Will its high concentration of ammonia upset the process?"; and "Which is the best proportion to mix both substrates?" The tests were performed in the author's lab, using an AMPTS II and digestate samples from the customer's plant as inoculum. Cow and rabbit manures were mixed in different proportions, and all were digested at the same temperature (38°C), which is a common practice to favor maximum biodiversity. The figure illustrates how the specific production of one of the mixtures yields significantly more methane than the single substrates considered individually.

Such a simple test showed the ideal mixture proportions to maximize the methane yield.

4.2.3 Preventing the Inhibition Caused by "Difficult" Substrates

In this context, a "difficult" substrate is any degradable substance that, in spite of its theoretical degradability, can upset the AD process because it contains some inhibiting compound. Some examples of such substrates are chicken dung (high ammonia and hydrogen sulfide), vegetable oil extraction byproducts (containing long-chain fatty acids, lecitin and waxes), olive mill wastewater (containing polyphenols), cheese whey and marine biomass (containing salt). Sometimes the concentration of inhibiting substances in the substrate would not pose any problem for a "healthy" inoculum, but in some biogas plants (especially in thermophilic ones), the bacterial ecosystem has little biodiversity and can consequently suffer a biological collapse when suddenly changing the diet.

The following is an extreme example that shows how a degradable substrate having a complex chemical composition can upset a biogas plant that "worked well" (but at the edge of the biological collapse, without the manager being aware). The biogas plant in question operates solely on corn silage, without the addition of any manure. The fermentation was induced and maintained in time by the regular addition of "a special formula covered by industrial secrecy" (presumably a mix of lyophilized bacteria and probiotic substances). The plant's owner pretended to replace part of the silage with slaughterhouse waste, composed mainly of lipids and proteins. Figure 4.5 very eloquently illustrates the disaster that the said sudden change in the digester's diet would have caused if no laboratory test had been performed in advance.

When sampling the inoculum, the plant manager reported: "It's all in order, because the FOS/TAC value is perfect". From the gas production curve *Acetate* (dashed line), we can conclude that the addition of acetate caused a partial inhibition of the inoculum, a fact demonstrated by 10 h of lag phase. It is possible to deduce from this test's results that the inoculum already contained a certain quantity of propionic and butyric acids, which the test FOS/TAC, performed by the biogas plant manager, measured as acetic acid. Furthermore, we must recall that this inoculum had a low buffer capacity

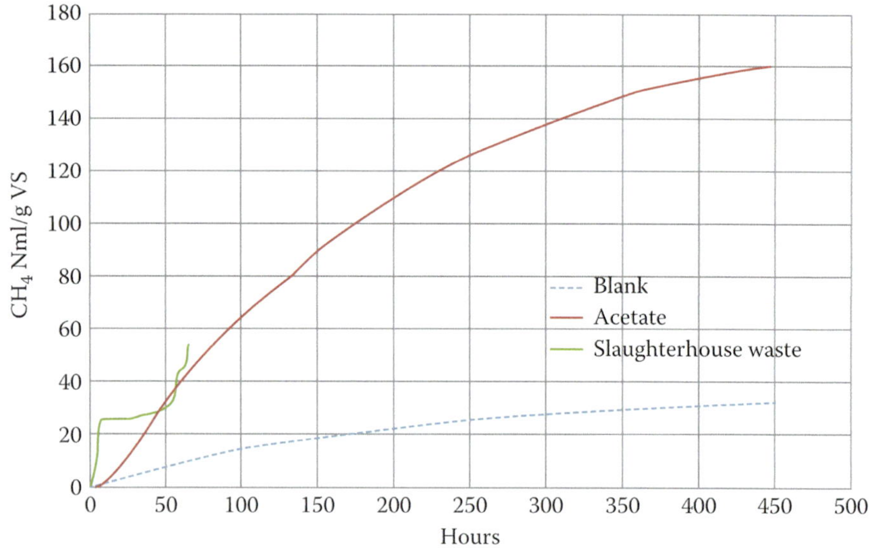

FIGURE 4.5
Example of inhibition caused by a "difficult" substrate because the inoculum's bacterial biodiversity and overall biological activity were scarce.

because the plant never received manure after its startup. Furthermore, degassing the inoculum before starting the test was deliberately omitted to simulate what would happen if the slaughterhouse waste were added to the digester. Adding a lipo-proteic substrate immediately produced a remarkable amount of methane during the first 10 h (probably because it created a favorable C/N ratio, or because the said substrate contained a small fraction of easily degradable amino acids). The surge of methane production was then followed by a short *plateau* (the beginning of the inhibition) and subsequently a short burst, which was suddenly interrupted (biological collapse).

4.2.4 Checking the Digestion Efficiency of the Biogas Plant

We have seen in Section 4.1.2 how to check the efficiency of the biogas plant in converting biomass into methane. Furthermore, it is crucial to evaluate the residual methanogenic potential of the digestate, to quantify its residual energetic value, and eventually to analyze possible strategies to extract the maximum profit from it.

The following example is based on a real case, involving a biogas plant with a single digester and an open tank for collecting the digestate. The digester was fed with a mixture of swine and bovine manure, corn, and triticale silages.

The VS concentration of the feedstock mixture sampled from the compensation tank placed before the digester was 8.75%, while that of the digestate, sampled at the digester's output, was 4.1%. The conversion efficiency in methane, according to the VS reduction criterion, was hence:

$$\eta_{DA} = \frac{8.75\% - 4.1\%}{8.75\%} = 53\% > 45\%$$

The plant's performance is good, although not at all brilliant. However, the formula of the VS reduction efficiency does not allow us to assess whether the residual fraction of the VS contained in the digestate is still capable of producing useful amounts of methane. To recover such residual methane, the simplest strategy possible in CSTR plants is just to increase the HRT. Hence, the digestate storage tank should be covered with an airtight dome and if possible heated too, to extend the digestion until the complete depletion of any degradable organic matter. For the plant owner, the important question was: "Is it profitable to invest money in covering the open digestate storage tank with a biogas collection dome?"

The biological test in such cases is very simple: it is enough to incubate the same quantity of sludge, sampled from the premix tank before the digester and from the digester's output, and calculate the fraction of residual BMP in the latter. Please note that a consistent percentage of residual BMP in the digestate, e.g., 20% of the feedstock's, not only means losing 20% profit, but it is also a source of greenhouse gas emissions to the atmosphere. Figure 4.6 illustrates the test's result.

It is quite evident from the curves that, in this case, the residual methanogenic potential of the digestate is still high, 50 Ncm^3/g VS, while the

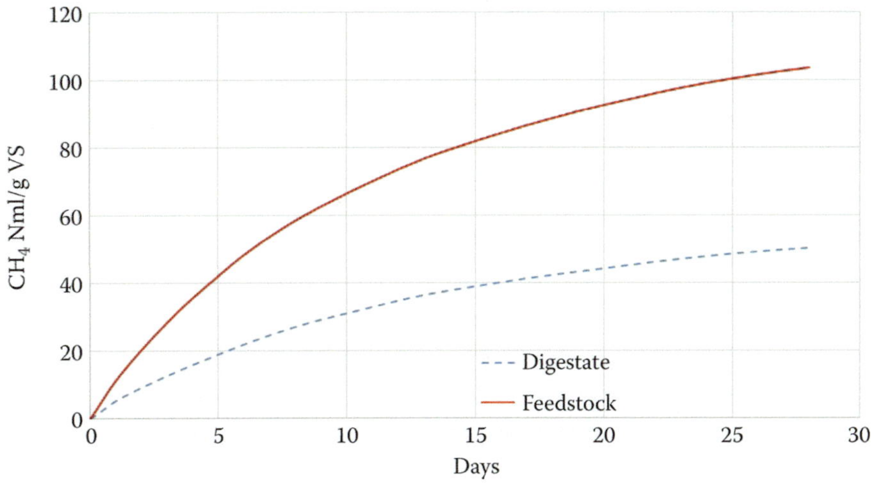

FIGURE 4.6
Residual methanogenic potential of the digestate, compared to that of the feedstock.

feedstock mixture has 103 Ncm^3/g VS. The resulting efficiency is hence 51%, a value coherent with the one estimated with the VS reduction criterion. Then, it is worth performing an economical evaluation of the investment in a biogas collection dome for covering the digestate collection tank. Such an investment will allow saving nearly 50% of the cost of silage for the rest of the plant's commercial life. Another option (not acceptable in most European countries because of environmental legislation) is to not cover the digestate tank and to replace the current silage with alternative substrates having a shorter digestion time (i.e., less cellulose). In such cases, there will be no investment cost, but the operational cost will be higher, since easily degradable biomasses (e.g., molasses, sugar beet pulp, alfalfa, cereal meals, etc.) are usually more expensive than corn and triticale silage.

4.2.5 Determination of the Optimum SRT/HRT

Most biogas plants in Europe are designed to operate with an SRT (solids retention time) ranging from 30 to 60 days, which is necessary for the complete anaerobic degradation of the most common substrates. In some cases, alternative substrates with short digestion times may be available (e.g., byproducts from the food industry, such as potato skins, or wastewater from beverage production). Figure 4.7 shows the digestion curves of wastewater from a beverage factory. The organic matter content of the said industrial effluent is very high: 69,600 mg COD/L. It is composed mainly of residual

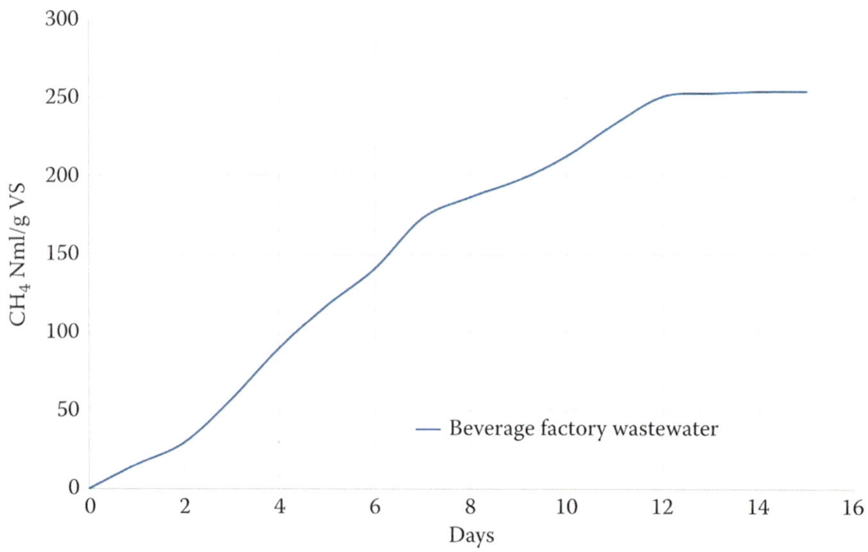

FIGURE 4.7
Anaerobic degradation of wastewater from a beverage industry.

starch and dissolved sugars, so their anaerobic degradation is quick, as clearly shown in Figure 4.7.

We can observe that the curve's shape has some "humps", but it remains almost linear until the complete depletion of the digestible fraction of organic matter on the 12th day. In that case, we can employ the liquid to dilute silage or any other solid feedstock composing the usual diet of the biogas plant, proportionally reducing their quantity. Caution must be taken to prevent the acidification of the digester. The conversion of sugars and starch into acetic, propionic, and butyric acids is faster than the conversion of said acids into methane. If the inoculum has enough alkalinity, or if manure is added as a co-substrate, there is little risk of acidification. However, in industrial plants that do not employ a high alkaline co-substrate, such as manure, it may be necessary to add alkalinity, typically as sodium carbonate or bicarbonate (at the risk of sodium causing inhibition), or lime (risk of sediments reducing the digestion volume), to neutralize the pH. In our example, the wastewater had pH = 6.4 when starting the test. Since the inoculum was sludge from an agricultural biogas plant running on cow manure, neutralizing the substrate before starting the test was not necessary.

4.2.6 Determining the Efficacy of Additives and Pretreatments

The use of commercial additives is probably one of the most discussed (and debatable) arguments when optimizing a biogas plant. The author's personal experience on this subject is that most of the arguments in favor of the said praxis are self-referenced pitches of the vendors or some user's belief that their plant will not work without the addition of probiotics, trace elements, desulfurants, etc. The vendors' arguments are usually weak and range from vague phrases such as "thousands of plants in Germany use our product" to "scientific studies made by the University of XY show that this product boosts the biogas production". Of course, upon reviewing the source, one finds that the "scientific study" was sponsored by the vendor, and in some cases, the results show a small improvement in total biogas production, without indication of the total error margin of the test or the net methane production. Some examples of said situations will be analyzed in Chapter 6. The factual reality is that the efficacy of a given additive varies from one biogas plant to another, or even over time in the same plant, depending on the initial state of the microbial consortium. In some rare cases, it is possible to observe miraculous performances; in most cases, nothing happens, and in some exceptional cases, the additive may even induce a slight inhibition. Figure 4.8 shows the effect of a product, containing live lyophilized bacteria, on an inoculum sampled from a biogas plant that had suffered a biological collapse. The reference substrate employed for the test was starch. We can observe that the reactors doped with the additive reached the theoretical BMP of starch in 10 days, while the control reactors, loaded with the same amount of starch but having the untreated inoculum, were not able to degrade the said simple substrate.

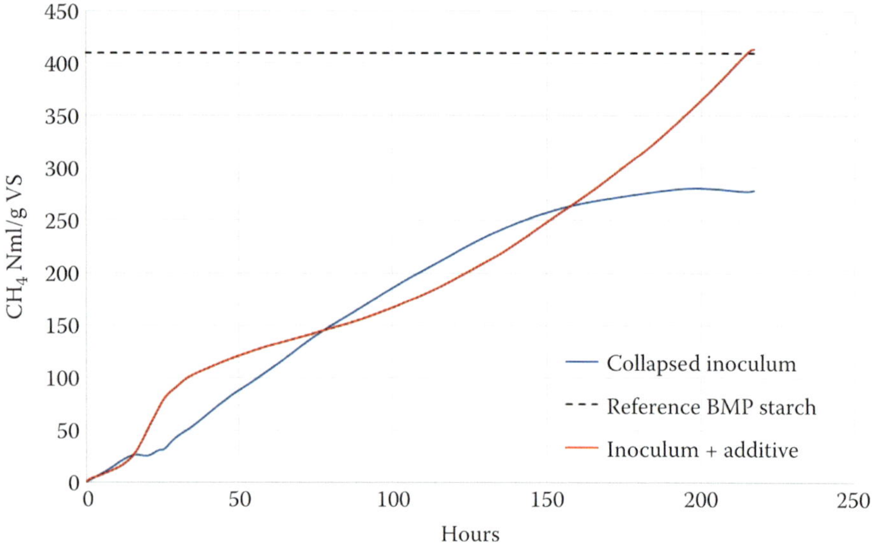

FIGURE 4.8
The effect of adding live bacteria to an inhibited inoculum.

The curve corresponding to the inoculum treated with the additive (red line) appears somehow "anomalous" (very different from the usual sigmoid). The cause is the sudden change in the composition and specific biological activity of the microbial ecosystem. Adding a new bacterial population to an inoculum that had already reached equilibrium caused a series of variations in the metabolic rate of the microbial system, until it reached a new state of equilibrium after 100 hours. From that moment onward, the methane production became constant (almost linear growth). On the other hand, the control reactors (continuous line), had their bacterial ecosystem already in equilibrium, although the biodiversity was poor. The result is a curve that is indeed a sigmoid, but very "flat" and having a lower asymptote than the starch's BMP. This means that the inoculum was not capable of degrading all the starch, even after 10 days.

Figure 4.9 shows the result of the digestion of shredded mixed waste from a pig slaughterhouse, employing a "healthy" inoculum. A comparative test was run, adding the same product of the former example to this inoculum and also boiling the waste but without adding any product to the inoculum. It is pretty evident that boiling lipoprotein-based substrates like the one in the example increases remarkably their BMP, while the difference between the digestion of the raw waste, with and without additive, is almost irrelevant (the difference between both curves lies within the uncertainty range of the test).

Another comparative test, this time between additive *A* (defined as a "probiotic mixture of alkaloids and trace elements") and additive *B* (defined

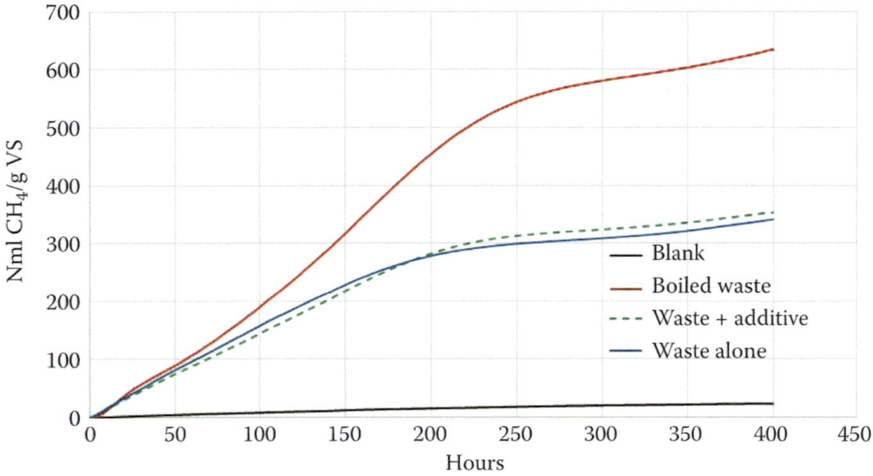

FIGURE 4.9
Comparison between the influence of a thermal pretreatment and the use of a commercial additive in the BMP of swine slaughterhouse waste.

as a "mixture of lyophilized bacteria, vitamins, and biocatalysts"), and the untreated inoculum, yielded the results shown in Figure 4.10.

Observing the curves, it is evident that additive *A* promotes the bacterial activity in an almost "miraculous" way. Nevertheless, the test showed that the concentration of methane diminished remarkably (below 50%!). This behavior is a clear demonstration that the additive has a positive effect on the fermentative bacteria, but little or no impact on the methanogenic flora. Furthermore, the curve of additive *A* shows "humps", while those corresponding to the blank and to additive *B* are almost perfect sigmoids, but "flat". Hence, it is necessary to thoroughly evaluate whether, from the operational point of view, the use of additive *A* is worth it. Figure 4.11 illustrates the net methane flow of the same reactors, i.e., the time derivative of the curves presented in Figure 4.10. In this case, the curves were obtained automatically using an AMPTS II, as it is a built-in function of the instrument. However, it would have been just as easy to produce them using a spreadsheet and daily readings from a manual instrument. Please note how additive *A* induces an irregular bacterial activity, showing remarkable production peaks followed by valleys (the "humps" in the curve in Figure 4.10, which may seem almost imperceptible to the unexperienced eye). Such an erratic production of gas is a drawback of additive *A*, as the flow peaks may lead to waste biogas in the plant's torch, while during the flow valleys, the amount of methane may be insufficient to feed the generator at nominal power.

In the case study above, the batch test provides valuable information to understand the benefits and drawbacks of additive *A*; however, it is not sufficient to define an operational strategy. The author's customer had requested

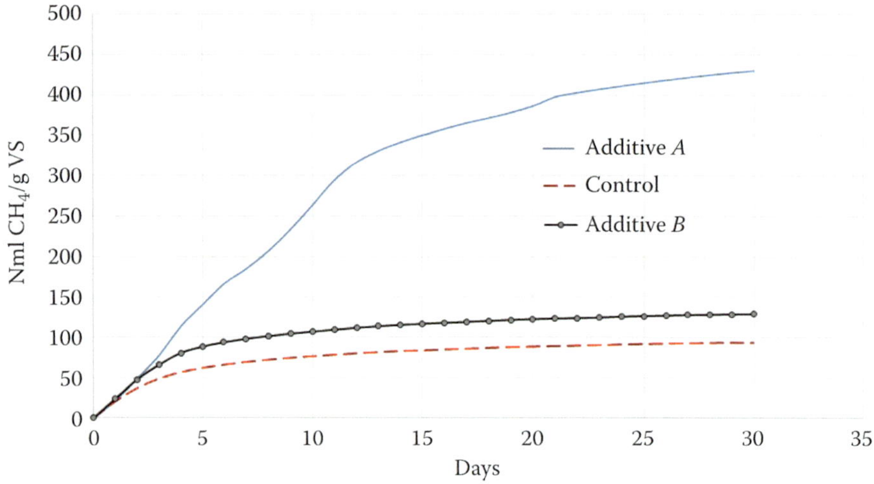

FIGURE 4.10
Comparison between the net methane yield of the untreated inoculum (control) and the same inoculum with two different commercial additives.

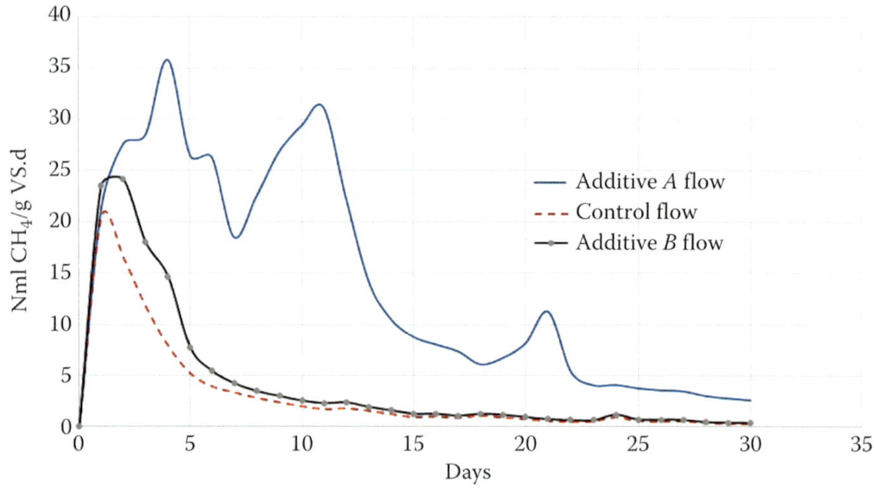

FIGURE 4.11
Net daily flow of methane produced by the control reactor and with two different additives.

to perform the test using the unfiltered inoculum, sampled from the digester, with and without additives. Such a procedure is not reliable, because the peaks observed in the batch test could be just the result of a transient situation, for instance, undigested organic matter accumulated in the inoculum that the additive made digestible. In such cases, it is more advisable to conduct a simulation of the process in a continuous reactor at different doses of additive *A*, as well as the simultaneous application of both additive *A* and *B*, to determine if a stationary methane production can be achieved. Nevertheless, continuous tests are more time- and labor-consuming than batch ones. A cheaper alternative solution is to conduct comparative batch tests using different reference substrates. The procedure should follow the regular setup: three samples of filtered inoculum should be treated with the additive, and three other samples should be left untreated. All six samples should be degassed before performing the test with the chosen reference substrate. The standard procedure mentioned above takes longer than the one selected for the experiment presented earlier, but leaves no room for doubt when analyzing the results.

Practical Conclusion

Employing commercial additives in a biogas plant does not necessarily improve the process. It is advisable to always measure in the lab whether adding a booster product to the inoculum is necessary, the optimum dose, and the eventual side effects (e.g., an increase in total biogas production accompanied by a reduction in the methane percentage). Once the objective data are obtained under controlled conditions, it is easy to plan a rational strategy to enhance the real-scale process or to refrain from spending money on unnecessary products. The same criterion applies to pretreatments. It is necessary to check whether they induce a beneficial or inhibitory effect, and if the application costs (both economic and thermodynamic, i.e., considering each electric kWh consumed in the pretreatment as roughly equivalent to 0.27 Nm^3 of methane) are worth the eventual benefits. Some additives contain heavy metals, albeit in minimal quantities. However, when the digestate is used as fertilizer, these pollutants end up in the soil and are absorbed by the crops, or may be leached into underground waters during rainfall. This environmental aspect must be carefully evaluated, without any prejudice, but based on objective figures.

4.2.7 Frequent Errors in Planning and Performing Biological Tests

Sometimes, the result of a biological test may cause perplexity, either because it is too different from expectations or because the specific methane production curves do not exhibit the classical sigmoid shape. The most frequent error is not degassing the inoculum sufficiently before starting the test. In such cases, the "background noise" induced by the inoculum, combined with some "difficult" substrate (e.g., containing large amounts of lipids or any inhibitor), will deform the curves of specific methane production. Figures 4.12 and 4.13 show examples of incorrectly performed tests. The inoculum was sampled from the digester and employed "as it was", without filtering and degassing it. The substrate to be tested contained a certain quantity of fats. Figure 4.12 shows how the curves of raw methane production look "quite normal", but

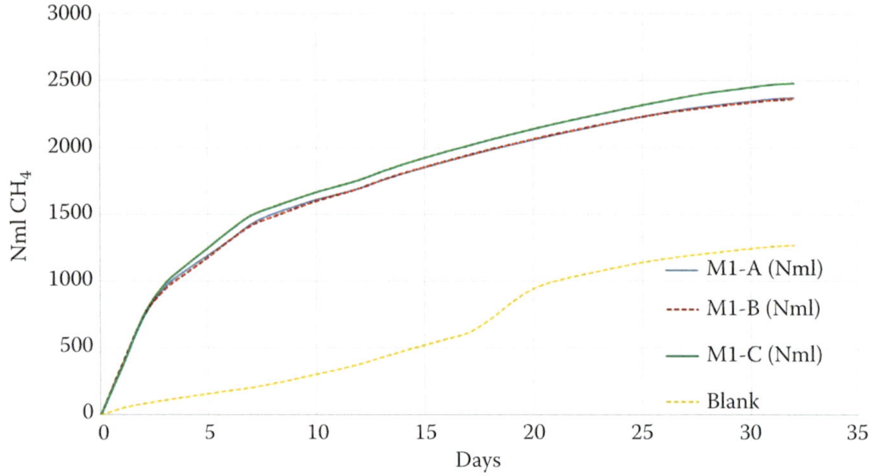

FIGURE 4.12
Individual CH_4 production curves of the blank and the sample reactors.

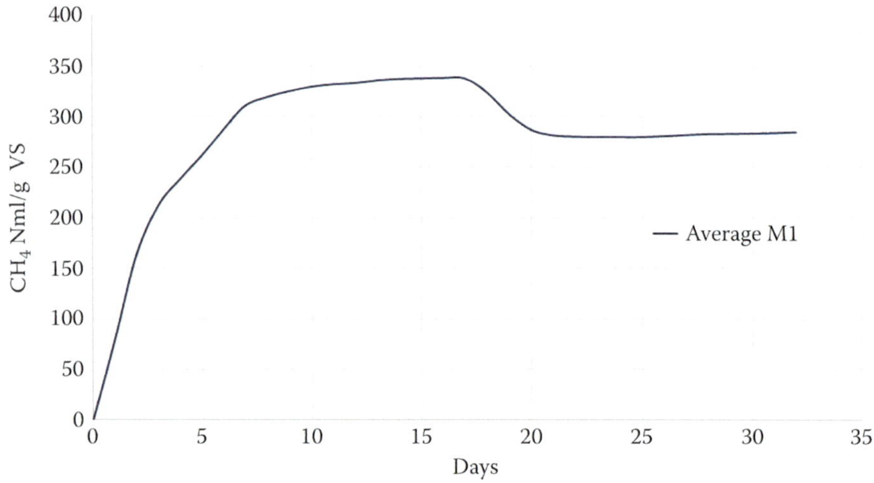

FIGURE 4.13
Example of an anomalous specific production curve.

after 30 days of incubation, they have not reached the plateau, which means that the substrate's digestion was only partial. Such behavior was expected, as lipids typically require 45 to 60 days for complete degradation. What the author did not foresee was the inoculum's behavior. In Figure 4.12, one can observe that between the 16th and the 21st day the curve has a "hump".

Probably, the cause of such anomalous behavior was the presence of some organic matter that was still in the hydrolysis and acidogenesis phase when the inoculum was sampled from the reactor.

This example featured partial inhibition caused by the fats and employed an inoculum without prior degassing (business as usual—the customer wanted quick results!). As was to be expected, when plotting the net CH_4 production curves (i.e., average of the samples' production minus the blank's production, divided by the VS), the result showed the anomalous shape that can be observed in Figure 4.13.

When the BMP test is performed correctly, i.e., thoroughly degassing the inoculum in advance, the specific production curve must constantly increase because, by definition, the curve represents the cumulative gas volume and hence cannot have a decreasing inflection, as shown in Figure 4.13. In this case, two factors contributed to yield a very anomalous specific production curve: having accepted the customer's request of employing a "fresh" inoculum (hence a high "background noise") and the partial inhibition caused by the fats. Both factors became evident from the 17th day, when the daily gas flow produced by the blank increased to a level higher than in previous days. At the same time, the sample reactors exhibited the opposite behavior, resulting in a decreasing net production curve. If the customer had not been in a hurry and the inoculum had been correctly degassed in advance, the curve would have been a perfect sigmoid, most probably showing a plateau from the 50th day, when the fats usually reach their complete degradation. In the present particular case, continuing the test up to 60 days—as established by the VDI 4630—had no practical sense, because the biogas plant in question was designed for a nominal SRT of just 30 days. Furthermore, the digestate collection tank was of the open type. In such particular cases, the plant manager must never forget that it is useless to buy a feedstock with a higher BMP (as all substrates containing lipids and proteins are) if the plant's SRT is shorter than the total time necessary for the total degradation of such substrate. In this particular example, the biogas plant manager had the opportunity to obtain such substrate at a favorable price if purchased before a specific offer expiry date, which was the reason for the urgency in performing the test and not degassing the inoculum. Consequently, the practical calculation of the digester's diet was based on the BMP measurement on the 15th day.

Important

The rules for planning and performing AD tests described in Section 3.8.1 are not just "academicians' fastidiousness". Not following the given procedures may lead to wasting time and obtaining useless results, as well as incurring additional costs and time spent on repeating the test. Remember, performing anaerobic digestion tests, even in the plant's laboratory, is easy and does not require a degree in chemistry. All you need is common sense, paying attention while preparing the test, and following the procedure without exception.

4.3 Using Reference Substrates to Check the Hydrolytic Activity: How to Find Out if Something Is Going Wrong

Together with the specific methanogenic activity (SMA) test, which we will study in detail in Section 4.4., it is advisable to perform the hydrolytic activity tests at least once a month, to assess if the bacterial ecosystem is healthy and to prevent the plant's biological collapse. We have already introduced the BMP of the reference substrates in Chapter 3, Table 3.3., and in Sections 3.10.1. through 3.10.5. We have described the special procedures and peculiarities of some substrates above, which require I/S ratios higher than 3 to prevent inhibition caused by volatile fatty acids (VFA), the presence of Na^+ ions, or long-chain fatty acids (LCFA). In the next example, taken from a real-life situation, we will learn how to interpret the test's results.

4.3.1 Hydrolysis Test of Cellulose

Cellulose-degrading bacteria thrive exclusively in the digestive tube of ruminants, specifically in their rumen. Other animals, including humans, are incapable of digesting cellulose, a fact exploited in some extreme weight loss diets, as eating cellulose in powder or toilet paper(!) fills the stomach and provides a sense of satiety, but does not provide nutrients. If, during the hydrolytic activity test of cellulose, the inoculum is not capable of reaching the reference BMP within 20–30 days, the biogas plant manager should immediately proceed to inoculate the digester (or the pre-fermenter, if the plant has two or more reactors) with cow, sheep, or goat manure. The inoculation

TABLE 4.3

Recipe of the Medium and Its Dosage to the Plant

	Compound Name	Concentration in the medium	1× compound concentration in the inoculum	Quantity of compound to the digester
Compound		**[g/L of Solution]**	**[g/m³ inoc.]**	**[g/1,000 m³]**
$MnCl_2 \cdot 4H_2O$	Manganese chloride tetrahydrate	0.05	0.5	500
H_3BO_3	Boric acid	0.005	0.05	50
$ZnCl_2$	Zinc chloride	0.005	0.05	50
$CuCl_2$	Copper chloride	0.003	0.03	30
$Na_2MoO_4 \cdot 2H_2O$	Disodium molybdate dihydrate	0.001	0.01	10
$CoCl_2 \cdot 6H_2O$	Cobalt chloride hexahydrate	0.1	1	1,000
$NiCl_2 \cdot 6H_2O$	Nickel chloride hexahydrate	0.01	0.1	100
Na_2SeO_3	Disodium selenite	0.005	0.05	50

will be even more effective if rumen can be obtained from a slaughterhouse. Some commercial probiotic products contain lyophilized bacteria; however, their high cost necessitates checking their efficacy in the laboratory before deciding to purchase them. If the hydrolytic activity test fails or shows a very slow degradation rate, then it is advisable to repeat it with different doses of rumen, commercial additives, or just manure. Bringing digestate from another biogas plant known to be "healthy" will usually help, but in this case, it is necessary to determine the minimum amount of additional inoculum required to reestablish an adequate bacterial population.

4.3.2 Hydrolysis Test of Sugar and Starch

This test can yield false-negative results if the inoculum has a low buffer capacity or if it already contains some amount of accumulated VFA deriving from other biological issues. Sugar and starch are simple, easily degradable carbohydrates. Their degradation produces alcohols, which in turn are further fermented into VFA. If the hydrolysis rate is higher than the rate of acetic acid conversion into methane, the reactor will tend to acidify. Hence, this test is complementary to the SMA test and more useful than the FOS/TAC (VFA/TA) test, providing a more comprehensive overview of the eventual imbalances in the sludge's bacterial ecosystem. It is customary in some countries to add alkalinity to the inoculum to neutralize the acidification caused by starchy or saccharine substrates (e.g., by adding sodium bicarbonate to the digester). Such practice only masks the symptoms without addressing the causes, and it must be avoided for two simple reasons: sodium bicarbonate is expensive, and the presence of additional Na ions can inhibit the process. The said argument will be analyzed in detail in Chapter 6. In agricultural biogas plants, there is a further reason to avoid using sodium bicarbonate: the presence of sodium in the digestate reduces its value as a fertilizer. It is preferable to check the effective hydrolytic activity in the laboratory and, if necessary, to add manure to the digester. Adding manure not only costs nothing in most cases, but it also provides enough alkalinity and additional living hydrolytic bacteria.

Practical Example

Figure 4.14 shows the anaerobic digestion curves of three products commonly available in supermarkets, which were used as reference substrates for testing the hydrolytic capacity of an inoculum sampled from a biogas plant that was not operating satisfactorily. It can be observed that their BMP is not much different from the theoretical BMP of the analogous pure laboratory reagents. Toilet paper tends to yield BMP values near the minimum because it is not composed entirely of cellulose. In this sense, natural cotton in flocks is better than toilet paper, since it is composed of 99% cellulose (Remark: Avoid cotton disks because they usually contain some wax). The shape of the anaerobic degradation curves of both starch and sugar does not follow the classical

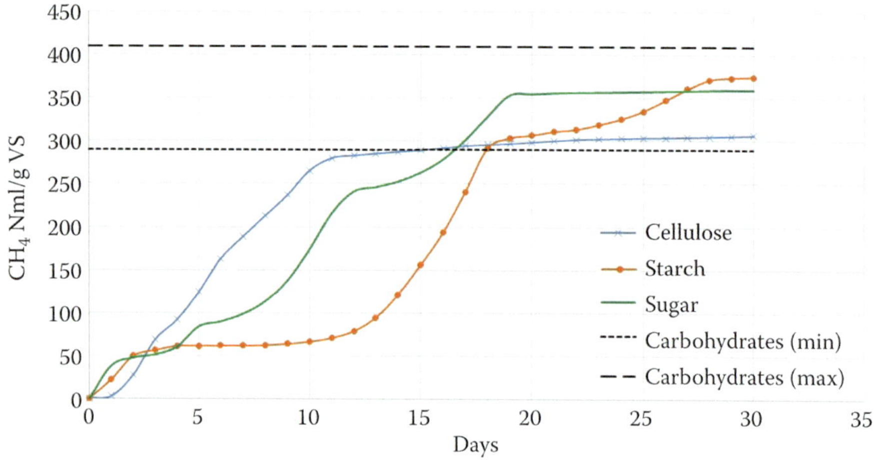

FIGURE 4.14
Anaerobic degradation curves of toilet paper (cellulose), white sugar (sucrose), and corn starch.

sigmoid pattern, but looks somewhat anomalous. It is impossible to establish *a priori* if such an anomalous shape is the result of the acidification caused by the high degradability of these substrates combined with a low buffer capacity of the inoculum, or if the latter has a low hydrolytic capacity, or if the real problem is a scarce methanogenic activity. The last supposition seems to be the most probable in this case because the degradation curve of cellulose (a more complex molecule, hence more difficult to digest) looks almost like a perfect sigmoid, although somehow "flat". All three reference substrates yielded an acceptable BMP, so it seems that the hydrolytic capacity of the inoculum can be considered "normal". To understand why the curves show such strange shapes, it is then necessary to compare the result of Figure 4.14 with the results of the additional tests that will be explained in the following paragraph.

4.3.3 Hydrolysis of Proteins

The hydrolysis of proteins is always slower than that of simple carbohydrates, but if the inoculum is "healthy", the complete degradation of protein should take about 10 days. The eventual preexisting accumulation of ammonia in the inoculum, plus the ammonia produced by the N contained in the proteins (typically 8% of the VS), can sometimes trigger inhibition phenomena. When encountering any issue with the proteolytic capacity test, it is necessary to check if the cause is inhibition by excess ammonia. A second test must be carried out, loading the reactor with the same total quantity of VS (same I/S = 3 as the proteolytic test), but in this case, half of the VS will be protein, and the other half will be cellulose. If the overall BMP is nearly the average of both individual BMPs, then the eventual problem is the accumulation of ammonia, rather than a lack of proteolytic capacity. If a photometer is available, measuring the ammonia content of the inoculum will be helpful, since concentrations above 3000 mg/L of ammonia are typically considered an alert limit by some authors. Nevertheless, the process temperature must also be considered, as the inhibition caused by ammonia is more likely in thermophilic plants and less troublesome in mesophilic ones. Should the ammonia be the cause of the low proteolytic activity, the solution consists of modifying the diet of the digester, reducing the percentage of protein- or ammonia-rich substrates (e.g., swine and chicken manure), and increasing the percentage of vegetal matter, which is usually richer in carbon than in nitrogen.

Should the ammonia levels be acceptable, then the scarce proteolytic activity can be improved by adding manure from monogastric animals (chickens, pigs, humans), which contains traces of proteolytic enzymes and a bacterial flora more suitable for hydrolyzing proteins compared to that of ruminants or herbivores in general.

Practical Example

Figure 4.15 presents the results of the proteolytic activity test, using fish gelatin as the reference substrate, with the same inoculum employed in the test shown in Figure 4.14. Once again, it can be observed that the inoculum, even reaching the minimum acceptable reference BMP, presents scarce specificity to digest protein. The shape of the curve is more regular compared to the curves produced by the digestion of sugar and starch. We cannot state *a priori* whether the bottleneck of the process is the proteolysis of the substrate or the final conversion step to methane. In this particular case, all points to the

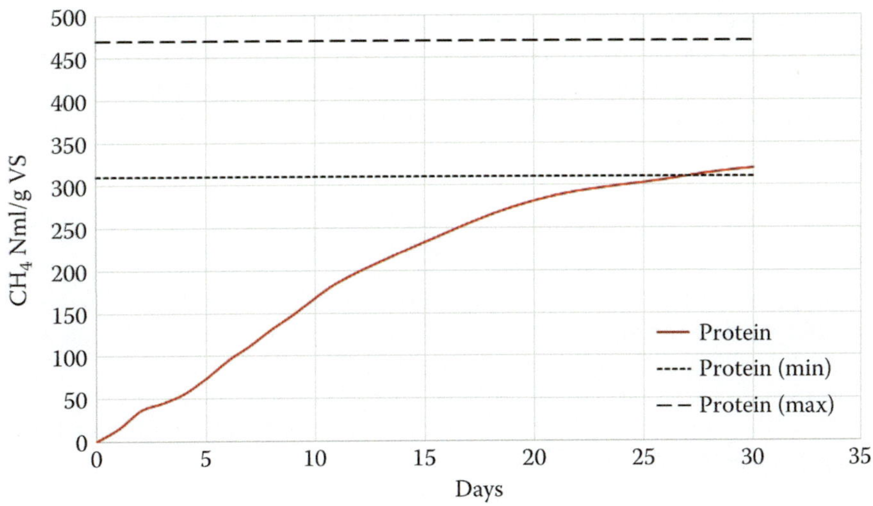

FIGURE 4.15
Proteolytic activity test using fish gelatin from the supermarket as the reference substrate.

second supposition, as already stated in Section 4.3.2, because the inoculum has a scarce SMA.

4.3.4 Lipolytic Activity Test

This test is performed seldom, but it must be considered mandatory if planning to feed the biogas plant with fatty biomasses (e.g., olive mill pomace, slaughterhouse waste, oilseed cake, kitchen waste, etc.). Always remember that if the plant is designed for HRT <60 days, attempting to feed it with fatty substrates will result in a failure, as the feedstock will not remain long enough in the reactor for its complete digestion. Furthermore, the presence of fats may hinder the complete degradation of other substances composing the diet. Should this test give negative results, the possible solutions are:

1 To inoculate the digester with swine manure.

2 To employ commercial additives containing lipase, which is the enzyme that hydrolyzes lipids, usually extracted from swine pancreas and the bile of any animal. In this case, the necessary dose may reach up to 1 g/L of pure enzyme to obtain a quick hydrolysis (4 h), resulting in a high operational cost. It is then necessary to perform laboratory tests, with different doses of enzyme, then check the minimum dose that allows complete degradation of the substrate within 15 or 20 days HRT, and finally to evaluate the operational cost and overall economic convenience.

3 To inoculate the digester with anaerobic sludge from a sewage treatment plant, if the local norms allow it.[2]

Alternatively, the slowest but the cheapest and surest way to employ fatty feedstock in a biogas plant consists of taking advantage of Darwinian selection. Perform the lipolytic activity test with several I/S ratios, find out the minimum one that does not inhibit the process, and start feeding the plant with such a dosage.

EXAMPLE

Suppose the digester has 1000 m^3 and its unfiltered sludge is 7% VS. After sieving the sludge, it has 3.5% VS. Suppose that the fatty feedstock is slaughterhouse waste containing 40% VS. Test the fatty feedstock with I/S = 3, I/S = 5, and I/S = 10. Suppose the tests show that I/S = 10 does not inhibit the process (perfect sigmoid shape, perhaps a bit flat). Since the amount of "live" sludge in the digester is:

$$I = 0.035 \cdot 1000 \text{ tons} = 35 \text{ tons VS}$$

The maximum daily dose of fatty feedstock should not exceed 3.5 tons VS/day, which means 8.75 tons of fresh slaughterhouse waste. As a precautionary measure, start feeding your plant with less than that, for instance, 7 tons/day.

N.B.: Fatty feedstocks usually have a long lag phase; hence, it is probable that during the first days, the biogas production drops. To avoid loss of production, it is possible to add some easily degradable feedstock during the first week, or start feeding a fraction of the maximum fresh feedstock dose and increase it gradually in the subsequent days.

After 1 month, the bacterial consortium will have adapted to the new diet containing fats, so the test should be repeated to measure the new lipolytic activity and maximum admissible dose. After a few months of gradual adaptation, the plant should be able to digest consistent proportions of fats in its diet.

4.3.5 General Substrate Inhibition Test

Sometimes it may happen that a batch of waste or byproduct from the food industry, or any other agricultural waste, is available at an attractive price for the biogas plant manager, potentially replacing the usual silage or commercial feedstock. The spontaneous question that the plant manager will pose to him-/herself is: "Will this stuff work? Won't it block the process?" Such questions are very pertinent, and one should always be aware that, if a batch of agroindustrial feedstock is very cheap or even free, the most frequent reason is that it is not apt for animal consumption or other uses. If a substance is toxic to an animal, it may also be harmful to microbes. Frequent examples include straw that caught rain and developed mold; meals and extracts from the oil industry treated with mineral acids or other chemicals; and slaughterhouse waste or byproducts not suitable for human consumption or pet food production. In other cases, the chemical features and consequently the BMP, together with the product's seasonality, make it impossible to define a stable market price (the typical case of olive pomace in the Mediterranean area). All mentioned byproducts are examples of potentially (but not necessarily) inhibiting substrates that can upset the anaerobic digestion process. In general, the anaerobic bacteria manage to digest even toxic substrates, but that may require longer times and lower concentrations than the usual feedstock to avoid triggering a biological collapse. In such cases, it is a good practice to measure the BMP, degassing the inoculum for at least a whole week, replicating the test with different I/S ratios, as explained in the former example. We can assume I/S = 10 as a practical limit, since a substrate capable of upsetting the process, even when fed in small amounts, poses more risks than benefits. The following example, taken from real life, shows the risks of adopting residual glycerol from the biodiesel industry as substrate for AD. The "literature" and "BMP databases" employed by biogas plant manufacturers and engineers give a misleading idea of said

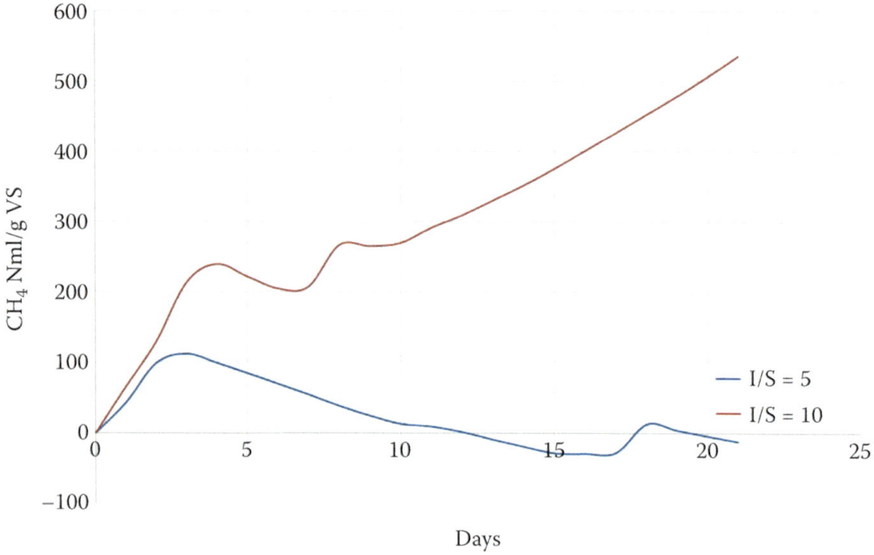

FIGURE 4.16
Inhibitory effect of residual glycerol on the AD process in a plant that has never employed such a substrate.

substrate, since its BMP is very high, its density is higher than that of water, and being liquid facilitates its handling. It may seem from such features that residual glycerol is the best AD feedstock one could imagine. What the "literature" does not say is that residual glycerol usually contains relevant quantities of soap, unsaponifiables, sodium chloride, or sodium sulfate, all of these substances with a certain inhibiting power for the archaea. Furthermore, the first step of the hydrolysis of glycerol is propionic acid, which we already know is a potent inhibitor of the methanogenic archaea. Figure 4.16 shows very eloquently the inhibiting effects of residual glycerol on the inoculum of a biogas plant running usually on corn silage and cow manure.

4.4 Applications of the SMA Test: Preventing the Biological Collapse and Selecting the Best Inoculum to Start a Biogas Plant

The theory and complexities of the SMA test, as well as a comprehensive method for the quick startup of digesters, are thoroughly explained in the book *Optimising Biogas Plants*, in this same collection. The present section will deal exclusively with the practical application of the SMA to the routine control of the digester's health.

The SMA test should be performed at least once a month, instead of the FOS/TAC (VFA/TA) test, since the latter test has scarce utility to prevent biological collapses. It is helpful to remember that the VDI4630:2016 norm recognizes that the FOS/TAC test is not selective, because it measures the mix of VFA present in the sludge *as if* they were pure acetic acid. The simplified SMA test proposed here (employing only acetic acid, or vinegar, or sodium acetate) is easy to perform, requires only 3 to 5 days to obtain a reliable result, and provides a clear insight into the bacterial ecosystem's health. Should propionic and butyric acids have accumulated in the sludge, as a result of some imbalance in the microbial population, adding acetic acid to it will provoke a partial inhibition of the acetoclastic archaea, resulting in an SMA value smaller than a given acceptable standard. Remember that the information obtained with the AD test of the acetic acid (or sodium acetate, or just ordinary vinegar) provides two useful figures assess the health state of the methanogenic archaea's: the SMA (quantitative, i.e., how active the archaea are) and the methanogenic capacity (qualitative, i.e., if there are active archaea in the inoculum).

4.4.1 Practical Example on How to Test an Inoculum Suspected of Methanogenic Inhibition Using Wine Vinegar

Figure 4.17 shows the result of a methanogenic capacity test performed with white wine vinegar as reference substrate, without neutralizing it. We can observe a perfect correspondence with the theoretical BMP of a 6% acetic

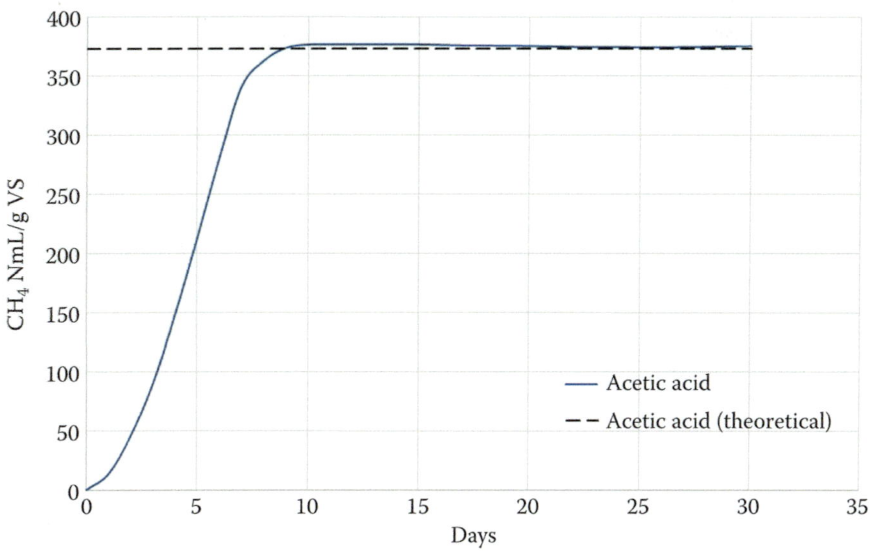

FIGURE 4.17
AD of white wine vinegar, employed as the reference substrate, compared to the theoretical BMP of a 6% acetic acid solution.

acid solution. This test was performed with I/S = 5. The inoculum was the same that was employed for the tests described in Figures 4.14 and 4.15. Comparing the different tests (sugar, starch, protein, and wine vinegar) from a superficial analysis, it may seem evident that the inoculum under test has indeed methanogenic capacity, its proteolytic capacity being rather mediocre, but acceptable. One could be tempted to assume that the hydrolytic activity is somehow compromised, since the curve in Figure 4.14 shows an unsteady behavior. Considering that the degradation of both protein and cellulose shows steadier dynamics, and that said substrates are more complex to digest than simple sugar and starch, we can deduce that the anomalous shape of the methane production curves of sugar and starch is the result of the hydrolysis speed being higher than the methanogenesis speed. This practical case demonstrates that checking just the methanogenic capacity of the inoculum is not enough. The methanogenic capacity is a qualitative test: It tells whether acetoclastic methanogenic archaea are present or not, by simply checking if acetic acid (or sodium acetate) produces methane and is completely degraded. The test does not say *how active or numerous* the acetoclastic archaea are. Thus, it is more useful to check the SMA, as will be described in the following paragraphs, since the figure gives *an objective measurement* of the archaea's population consistency and activity.

Figure 4.18 shows the result of the SMA test. It employs the same input and experimental data with which Figure 4.17 was plotted, but calculated differently. To generate the SMA curve, the experimental input data is the daily flow of methane produced by the acetic acid (or acetate), instead of the

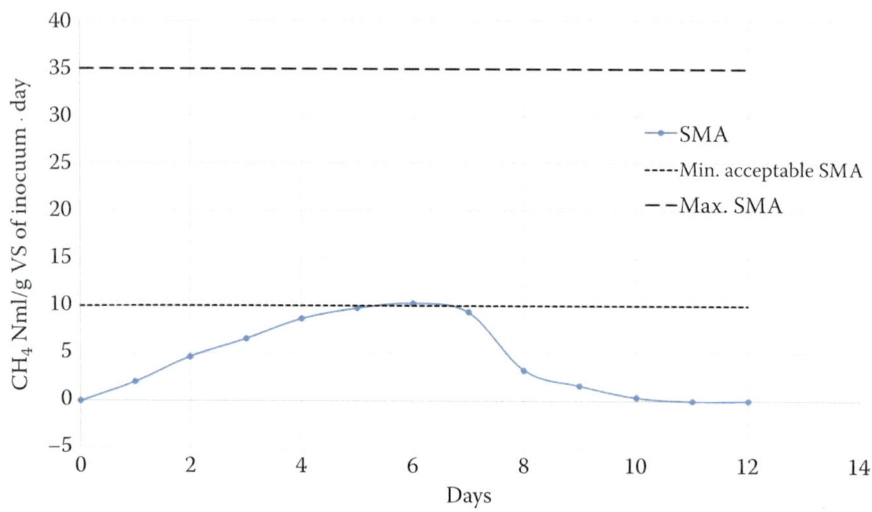

FIGURE 4.18
SMA test performed with wine vinegar as the reference substrate and the same inoculum employed for the hydrolytic capacity tests shown in Figures 4.14, 4.15, and 4.17.

cumulated volume of methane. If employing an AMPTS, the calculation is straightforward because the instrument provides reports in Excel format of both the cumulated volume and daily flow of each reactor. Other instruments may require a bit more manual calculations, but the eventual additional effort is irrelevant. The general idea of the SMA test is similar to that of the BMP test: Subtracting the daily methane production of the blank reactor from that of the reference reactor introduces the variable *time*, i.e., one has a clearer idea of *how quickly* acetate is converted to methane. Since the VSs of the inoculum are assumed to be a representative measurement of the amount of living bacteria, dividing the daily methane flow by the VS of the inoculum gives a figure that shows *how active* the archaea population is. When plotting the result of said calculations, the curve usually has the shape of a bell. By definition, the maximum value (the peak of the bell) is the SMA.

In the literature, there is a significant disparity in the criteria for calculating the peak SMA value. Some researchers check the hourly methane production rate and extrapolate to 1 day the maximum hourly value (i.e., multiply by 24 the measured value, resulting in very high numerical coefficients). A few other researchers employ a mixture of acetic, propionic, and butyric acids (or their sodium salts) as reference substrate, which is scientifically more correct than using only acetic acid, but makes the test more challenging to perform in an industrial plant's laboratory. For the practical purpose of managing a biogas plant, we will always perform the simplified test (with acetic acid, sodium acetate, or just wine vinegar), adopting the net daily methane flow as input data for the calculation. Under said rules, the acceptable values of the SMA will be in the range of 10–30 $Ncm^3/g{\cdot}VS_{inoc}{\cdot}day$.

Returning to the example of the inoculum with inhibition problems, when plotting the SMA curve, it is evident that the inoculum in question has an SMA value, which is at the lowest acceptable limit. This explains the anomalous shape of the curves in Figure 4.13. We can hence conclude that, in the case under study, there is no problem with the hydrolytic capacity; on the contrary, it is higher than the methanogenic capacity; hence sugars and starch are converted into VFA at a higher rate than the archaea are capable of metabolizing into methane, resulting in VFA accumulation. The acidification of the inoculum leads to partial inhibition of the archaea, which explains why the methane production curves of sugar and starch present "humps". Protein and cellulose, being complex substrates, have slower hydrolysis rates, so VFA are produced more or less at the same speed at which the archaea can convert them into methane, resulting in regular methane production curves (although a bit "flat" for a trained eye).

In the case of the biogas plant where the situation described was found, the owner had never been able to explain why the plant could run acceptably (although not without some troubles) when fed with corn and triticale silage, while losing production when fed with sugar beet or molasses. The owner had spent a lot of money on buying a titrator and trying to manage the process through monitoring the FOS/TAC (VFA/TA) parameter, which showed no anomalies because the inoculum's high alkalinity (dry cow manure) masked the acidification. Sending samples to an external laboratory to analyze the VFA profile had proved of no help to solve the enigma, until the simple tests with reference substrates described earlier provided a clear vision of where the biological bottleneck was.

4.4.2 Practical Example of the SMA Test for the Selection of the Inoculum Necessary for Starting a New Biogas Plant

Another application of the SMA test is the selection of the most suitable inoculum for starting a new biogas plant (or for restarting a plant after a biological collapse, or for "reinforcing" the bacterial ecosystem in a troublesome plant, such as the one described in the former example). The following example, taken from a real case, shows how the inocula from different biogas plants may show very different SMAs, despite all of them being apparently stable in operation. In this case, the mission was to start a newly built agricultural biogas plant. The options to inoculate it were:

1. Taking sludge from a nearby biogas plant that treated industrial effluents (inoculum A).
2. Taking sludge from an agricultural biogas plant fed with cow manure and corn silage (inoculum B).
3. Taking sludge from a biogas plant fed mainly with mixed silage, with cow and pig manure added from time to time (inoculum C).

The corresponding SMA curves are shown in Figure 4.19.

All three inocula reach peaks higher than the minimum acceptable SMA (10 $Ncm^3/g{\cdot}VS_{inoc}{\cdot}day$); nevertheless, inoculum B appears to be the best one. The negative portion of inoculum B's curve is the result of a "quick test", i.e., the inoculum had not been degassed enough. Inoculum C is good, while inoculum A probably has some deficit of trace elements, because the SMA curve shows some ripple and reaches the peak a bit "late", on the 5th day.

4.4.3 How to Determine the Dosage of Trace Elements if SMA < 10

The same basic SMA test already described will be repeated, but this time with three additional reactors, each one added with increasing dosages of *medium*. Medium, a.k.a. *mother solution* or *micronutrients solution*, is an aqueous solution containing trace elements at a concentration 1000 times higher than the minimum, so that adding 1 mL of medium per liter of inoculum should provide the acetoclastic archaea with the minerals they need. Adding different dosages to the reactors will allow comparing the SMA of each one, and finding out the optimum, which may be higher than the minimum threshold. A recipe of the medium, according to the Italian UNI/TS11703:2018, is provided in Table 4.3. The reader may employ other recipes found in the

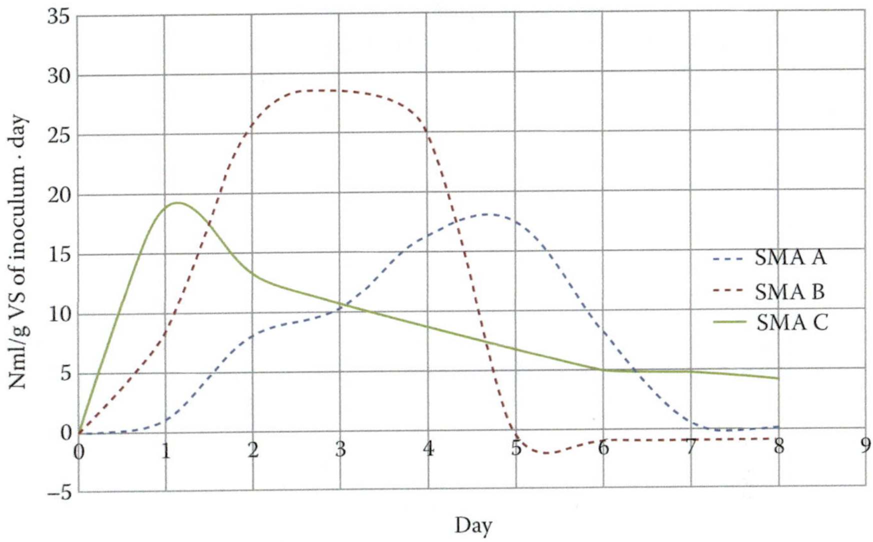

FIGURE 4.19
Different SMA curves generated by inocula sampled from biogas plants running on different substrates.

literature, paying attention to the dosage specified in that case. The steps for performing the test are the following:

1. Sample the inoculum from the digester and sieve it through a 5-mm mesh. Measure its VS or its COD.
2. Fill four reactors with the same volume of inoculum, *V*, and label them "control", "1×", "2×", and "3×". Add 10 mL of medium per liter of inoculum to "1×", 20 mL/L to "2×", and 30 mL/L to "3×". A micropipette or a burette will be very useful for correctly dosing the medium into each reactor.
3. Close the reactors, flush the headspace, and preincubate them for at least 7 days at the same temperature as your plant. The scope of this operation is double: degassing the inoculum for higher test accuracy and allowing the micronutrients to be absorbed by the microbes, thus reactivating their enzymatic activity. It is advisable to check the biogas production until the daily specific methane production is in the range 2–5 Ncm^3/g VS of inoculum·day.
4. Once the four reactors are degassed, add to each one 33 g of vinegar per liter of inoculum. It is not necessary to open the reactors—the advantage of using vinegar instead of sodium acetate is that the former can be added with a syringe through the tube with valve used for flushing the headspace volume at the beginning of the test, without opening the reactor.

5 Start the experiment, wait at least 4 days, and check the daily flows to find the maximum daily biogas production of each reactor—in general, it will be reached within 48–72 h. We call this value "Peak" [Ncm^3/day].

6 Calculate the SMA of each reactor with the following formula:

$$SMA_i = \frac{Peak_i}{V \cdot VS_{inoculum}}$$

The micronutrient dosage for the plant will be calculated from the results of the test described earlier. The ideal micronutrient treatment is the minimum dosage that produces the maximum effect. For instance, suppose that the calculated SMAs are the following:

$$SMA_{control} = 6\ Ncm^3/g\ VS \cdot day$$
$$SMA_1 = 11\ Ncm^3/g\ VS \cdot day$$
$$SMA_2 = 20\ Ncm^3/g \cdot VS \cdot day$$
$$SMA_3 = 23\ Ncm^3/g \cdot VS \cdot day$$

The $SMA_{control}$ value of this example is typical of a plant suffering biological collapse. The dosage 1× could suffice to bring it to a minimum activity, but the resulting SMA is just slightly higher than the acceptable minimum. The dosage 2× brings the plant to a "good" SMA. Dosage 3× produces minimal improvement compared to dosage 2×, so its application would be a waste of money on overdosing. Suppose our digester has 5000 m^3. The required quantity of each micronutrient compound will be 5 times the figure in the fifth column, multiplied by 2 because the optimum dosage is 2×, i.e., 5,000 g of $MnCl_2 \cdot 4H_2O$, 500 g of H_3BO_3, and so on.

Figure 4.20 shows the result of adding the calculated dosage to a 1-MW (electric) biogas plant that was inhibited by the lack of trace elements. The increase in output power was relatively steady after 5 days of the treatment.

4.5 Conclusions

4.5.1 Usefulness of the Test with Vinegar, Acetic Acid, or Acetate as Reference Substrate

The BMP test using vinegar, acetic acid, sodium acetate, or a mixture of VFA as the substrate provides two pieces of information that the biogas plant manager must evaluate *simultaneously*. The first is the *capacity* to convert all the substrate into methane (hence, the ultimate methane production compared to the theoretical BMP of VFAs). The second is the *maximum speed* at which the said conversion is achieved, referred to as the mass of living bacteria

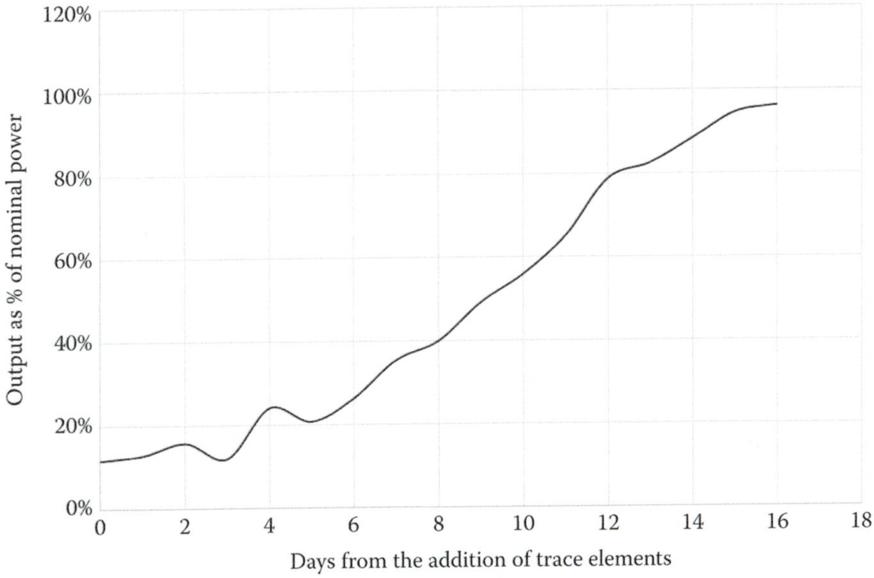

FIGURE 4.20
Effect of adding the correct dosage of trace elements to a biogas plant that presented impaired methanogenic activity.

represented by the VS (i.e., the SMA). The SMA is useful both for periodically checking the health status of the digester's bacterial ecosystem and for selecting the most suitable inoculum to start a biogas plant.

4.5.2 Reference Substrates

The practical examples presented in Sections 4.3 and 4.4.1 demonstrate that some simple tests performed with cheap and widely available reference substrates provided more helpful information than that obtained through expensive analysis performed by external "specialized" laboratories.

In conclusion, it is more important to know *how to* measure the real microbial activity and digestion efficiency, rather than deciding our plant management strategies on generic guides, tables from "the literature", and expensive tests performed by third parties.

Notes

1 Author's note: Most biogas plants in Italy still sell only electricity. Some amount of heat is employed to keep the digester's temperature, and the excess heat sometimes has marginal uses for greenhouse heating. In our example, excess heat is just dissipated to the atmosphere, hence in this calculation the economic value of heat is null.

2 Note of the author: in Italy and some other European countries, using sewage sludge in an agricultural biogas plant may represent a legal problem, because it is forbidden to employ the resulting digestate as a fertilizer. From a scientific point of view, such regulations are absurd and even contradictory to the EU Directives on circular economy and "end of waste status". On the opposite side, the Swedish regulation *SPCR 120—Certification rules for digestate from biowaste by the quality assurance system of Swedish Waste Management (December 2007)* is probably the most pragmatic and scientifically rigorous example of digestate management policy in Europe. According to the said regulation, the acceptability of the digestate for agriculture depends just on its quality, evaluated through well-defined chemical analysis. Anyway, the biogas plant manager must be aware of the applicable local regulations before making any decision.

Bibliography

Bolzonella, D., *Monitoring and Lab Analysis—Assays, Notes on the Course on Biogas Technology.* Jyväskylä Summer School, Jyväskylä, Sweden, 2013.

Burke, D., *Dairy Waste Anaerobic Digestion Handbook.* Environmental Energy Company, Olympia, WA, 2001.

Hutòan, M., Mrafkovà, L., Drtil, M., and Derco, J., Methanogenic and nonmethanogenic activity of granulated sludge in anaerobic baffled reactor, *Chemical Papers, 53*(6), 374–378, 1999.

Lozano, C.J., Mendoza, M.V., de Arango, M.C., and Monroy, E.F., Microbiological characterization and specific methanogenic activity of anaerobe sludge used in urban solid waste treatment, *Waste Management, 29*, 704–711, 2009.

Miranda, L., Henriques, J., and Monteggia, L., A full scale UASB reactor for treatment of pig and cattle slaughterhouse wastewater with a high oil and grease content, *Brazilian Journal of Chemical Engineering, 22*(4), 601–610, 2005.

Rosato, M.A., *Quick restart of a biogas plant and micronutrients cost minimization, Application note of Bioprocess Control AB*, 2016. www.bioprocesscontrol.com/company/media-gallery/application-note-quick-restart-of-a-biogas-plant-and-micronutrients-cost-minimization/.

Rosato, M.A., *Optimising Biogas Plants – Squeezing Profit Out of Your Digester.* CRC Press, UK, 2025. https://www.routledge.com/Optimising-Biogas-Plants-Squeezing-Profit-Out-of-Your-Digester/Rosato/p/book/978104100329

Schneiders, D., da Silva, J.D., Till, A., Lapa, K.R., and Pinheiro, A., Atividade metanogênica específica (AME) de lodos industriais provenientes do tratamento biológico aeróbio e anaeróbio, *Revista Ambiente & Água—An Interdisciplinary Journal of Applied Science, 8*(2), 135, 2013.

Van Lier, J., Sanz, M., and Lettinga, G., Effect of temperature in the anaerobic thermophillic conversion of granular and dispersed sludge, *Water Research, 30*, 199–207, 1996.

Zhao, Y., Ma, S., Sun, Y., Huang, Y., and Deng, Y., Isolation, identification and enzyme characterization of a thermophilic cellulolytic anaerobic bacterium, *Wei Sheng Wu Xue Bao, 52*(9), 1160–1166, 2012 (in Chinese).

5

Some Simple Tricks to Improve the Laboratory's Operability

5.1 Foreword

Most of the techniques illustrated in this chapter were implemented and fine-tuned with an AMPTS II instrument, and the photos show the same, just because the author's laboratory has one. Nothing prevents that, and with the necessary adaptations, they can be implemented with other instruments too.

5.2 Measuring the Percentage of Methane in the Biogas with the Double Reactor Set and with the Syringe Method

In the academic world, the most widely diffused technique to analyze the chemical composition of biogas is gas chromatography. The said method is very accurate, but its cost and complexity place it beyond the possibilities of the average biogas plant manager. Almost all industrial biogas plants are equipped with a fixed online analyzer of the biogas composition, usually based on the infrared spectrometry (IRS) method to determine the proportions of carbon dioxide and methane. Portable biogas analyzers with IRS and/or solid-state sensors are also available on the market. Such instruments are not suitable for laboratory measurement, since they require "big" volumes of biogas at each reading. For instance, the Geotech Biogas 5000 portable biogas analyzer requires a biogas flow of 500 mL/min, far beyond the capacity of any laboratory reactor. Sometimes, the researcher desires to check in the laboratory not only the net amount of methane produced by the samples, but also the quality of the biogas, i.e., its methane percentage and content of trace gases. A simple way to obtain two individual curves, one showing the production of methane and the other the production of carbon dioxide, consists in measuring the difference between the daily production of biogas and the daily production of methane. To perform this test with an AMPTS II,

DOI: 10.1201/9781003676201-5

two identical sets of blank and sample reactors must be prepared. The first one must be connected to the instrument as usual, i.e., through the caustic soda filters, measuring the net methane production. The second set must be connected directly to the instrument, measuring the total biogas production.

With the AMPTS III series, there are two alternatives to perform the said test: using the AMPTS III Duo or using two cells and a CO_2 trap connected in series, for each reactor of the AMPTS III Standard. The peculiarities of the AMTS III Duo and the error propagation analysis of this particular configuration are explained in detail in the book by the same author, *Optimising Biogas Plants*. When using an AMPTS III Standard for this specific test, only nine reactors will be available because the instrument has eighteen cells. The elaboration of the data with the spreadsheet will be the usual—subtract the average of the blank reactors from the average of the corresponding sample reactors, and divide the said difference by the VS of the substrate. We will then obtain two columns in the spreadsheet, one containing the average daily specific production of biogas, and the other containing the average daily specific methane production. It is easy to add a third column to the spreadsheet, containing in each cell the difference between the specific biogas production and the specific methane production. The same represents the specific carbon dioxide production. Now it is easy to plot both curves, net methane and carbon dioxide productions, as a function of time. Alternatively, one can divide the cells containing the average methane production by those containing the average biogas production, and plot the percentage of methane as a function of time. Figure 5.1 shows an example of the results obtained with the test described here.

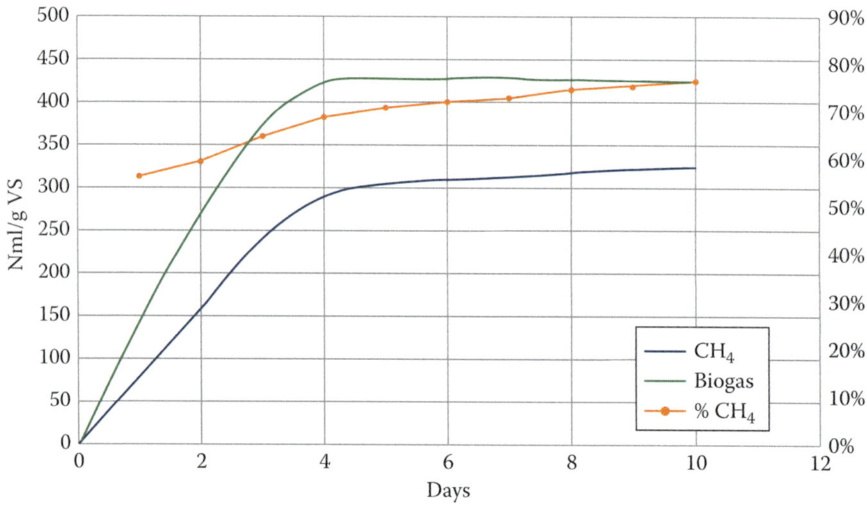

FIGURE 5.1
Comparison between the normalized curves of biogas and methane production, and the variation of methane percentage in the gas as a function of time.

Observe that the CH_4 percentage measured in batch tests like the one shown in Figure 5.1 tends to be overestimated in comparison with real-life experience with the same substrates, especially at the end of the test. The reason is that not all the biogas produced by the inoculum is released and passes through the instrument. Indeed, some of the CO_2 is captured by the inoculum itself, forming carbonates and bicarbonates. In some extreme cases, for instance, in the SMA test using sodium acetate, the biogas composition resulting from the batch digestion may reach 90% of CH_4, because the CO_2 is immediately captured by the NaOH that forms when the free Na, released by the archaea while digesting the $C_2H_3NaO_2$, reacts with the surrounding water and then with the CO_2. This is a feature of the batch tests, not a defect of the instrument or an error while performing the test. The only way to minimize the said effect would be to saturate the sludge with CO_2 before starting, but such an operation adds complexity to the test protocol and may alter the inoculum's buffer capacity. Figure 5.2 shows an example of the double reactor test with AMPTS II, using sodium acetate as substrate.

The comparison of the three curves in the same graph provides valuable information about the kinetics of the substrate's degradation, since the proportions of methane and carbon dioxide vary with time according to the fermentation rate, and this factor should be considered in some plants, especially those with two or more stages. For sure, performing two sets of the same test in parallel, with and without carbon dioxide filters, is the most complete method to understand the dynamics of the AD process, since it provides *continuous* information about the hydrolysis and methanogenesis rates. On the other hand, such a method requires employing twice the number of reactors

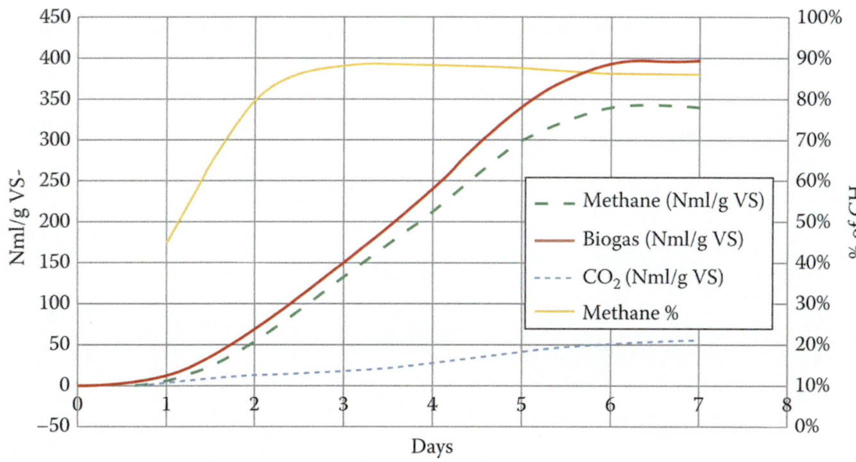

FIGURE 5.2
Specific production of raw biogas and net methane from sodium acetate, as measured with the double batch test. In this particular case, the percentage of methane in biogas can reach beyond 90%. The theoretical methane concentration from acetate digestion is 100% because the sodium contained in the acetate immediately captures the carbon dioxide as it is formed.

for the same substrate. Users who do not possess an AMPTS II or similar, i.e., an instrument having many measurement channels with real-time data logging and normalization, will have to perform more manual reading, normalization, and calculation to generate the curves described. In short, measuring in parallel total biogas and net methane requires some extra work. Additionally, this variation of the BMP test requires occupying reactors that we could otherwise employ for testing more substrates or for other biological activity tests.

The syringe method, described in the following paragraph, is straightforward and acceptably accurate for discrete measurements (e.g., once a day) of the biogas composition, assuming the latter as a binary mixture of methane and carbon dioxide. It can be employed for both batch and continuous tests, since it provides just points, for instance, one measure per day. The method is based on the extraordinary capacity of aqueous caustic soda solution to absorb carbon dioxide, and on *Avogadro's Law*:

> The volume occupied by a mixture of two gases is the sum of the volumes occupied by the single gases under the same conditions of pressure and temperature.

The validity of this method is justified because the other components of biogas (ammonia and hydrogen sulfide) exist in concentrations of the order of parts per million, negligible if compared to the concentrations of methane and carbon dioxide, which are always around 50%.

5.2.1 Necessary Materials for the Syringe Test

- A special syringe for gas sampling, at least 15 mL capacity, with valve (Figure 5.3)

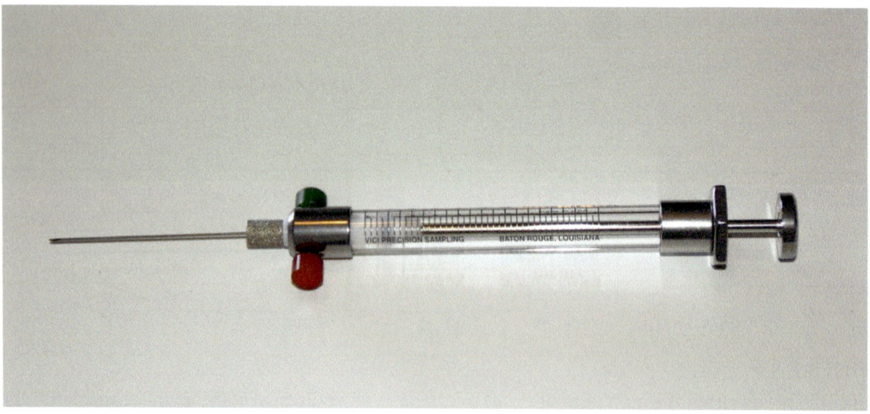

FIGURE 5.3
A special syringe for gas sampling, equipped with valve. (Photo by the author, 2-mL model manufactured by VICI, www.vici.com/syr/a2.php.)

FIGURE 5.4
Gas sampling valve with rubber septum. Courtesy of BPC Instruments AB.

- 1 g of caustic soda diluted in 10 mL of distilled water
- A gas sampling valve with rubber septum (Figure 5.4)

5.2.2 Performing the Test with the Syringe

1. Insert the needle in the septum, open the valve, and aspirate 10 mL (or 1 mL, depending on the available syringe) of biogas. Wait a few seconds until the pressure reaches the equilibrium and close the valve. See Figure 5.5.
2. Dip the needle in a small recipient containing caustic soda solution, open the valve, and aspirate at least 10 mL (or 1 mL in this example) of solution. Close the valve and shake the syringe for a few seconds. The caustic soda will absorb the CO_2. See Figure 5.6.
3. Immerge again the needle in the caustic soda solution, open the valve, and push the plunger to its initial position (10 or 1 mL in this example). Do not close the valve, and allow some seconds for the pressure to stabilize.
4. While keeping the needle immersed in the solution, hold the syringe in a vertical position and read the remaining volume of solution; see Figure 5.7.

The said volume of aqueous solution corresponds to the initial volume of carbon dioxide present in the biogas sampled. Conversely, the remaining volume of the syringe is now full of methane. In the example shown in Figure 5.7, the volume of solution is 0.30 mL; consequently, the biogas sampled from the reactor is composed of 30% carbon dioxide and 70% methane.

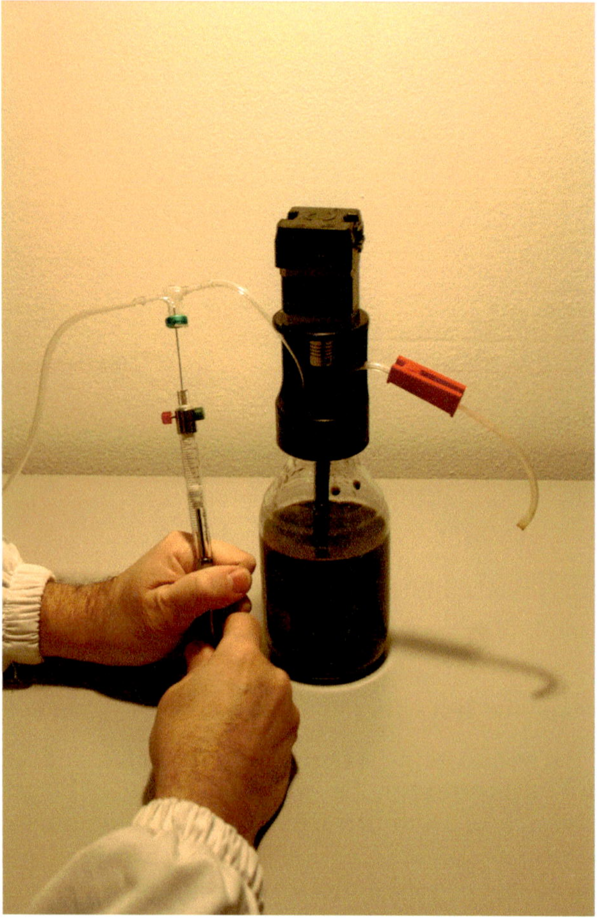

FIGURE 5.5
Aspirating the biogas sample. The green button shifted to the syringe body indicates the position "open." Press the red button before retiring the syringe from the sampling septum.

5.2.3 Error Analysis of the Syringe Method

As in all volume measurement methods, the error sources are:

1. The error class of the syringe. The minimum reading (absolute error) of a well-trained eye is half the space between two divisions of the volume scale engraved on the syringe. In our example, the scale spacing is 0.5 mL, hence the absolute error of the reading is ±0.25 mL, corresponding to ±2.5% absolute error of the CO_2 percentage.
2. Human errors. The superficial tension of water surfaces forms a meniscus. When reading the volume of solution remaining in the

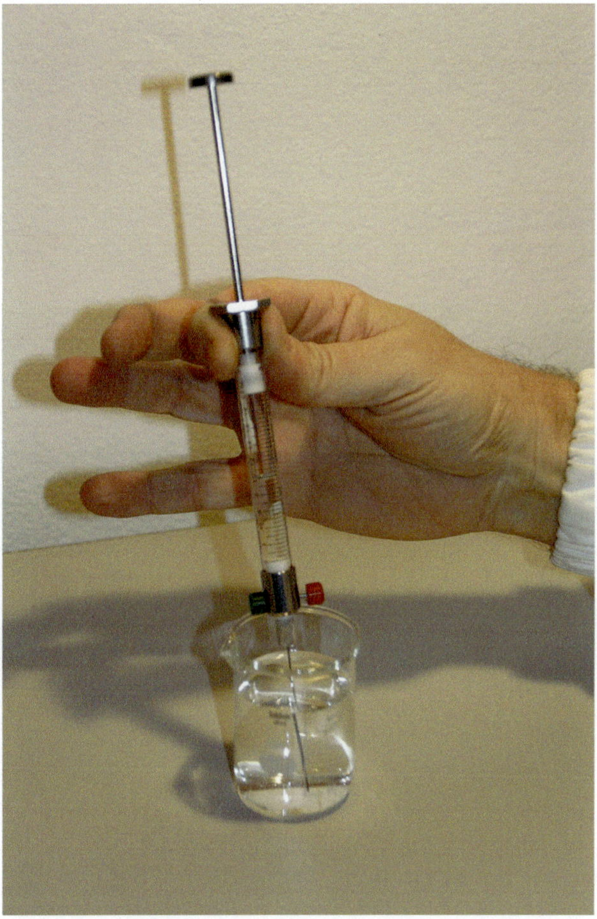

FIGURE 5.6
Syringe full of biogas and caustic soda solution.

syringe, the latter must be kept perfectly vertical, and the reference line is the lowest point of the meniscus.

Any of the following strategies can improve the accuracy of the syringe method:

1 Sample a bigger biogas volume (e.g., employing a 50- or 100-mL syringe). This is feasible only if the reactor's head volume is big enough. A 500-mL reactor, as the one shown in Figure 5.5, does not allow sampling more than 10 mL of biogas; otherwise, the internal pressure would drop to the point of saturating the headspace with water vapor. The vapor would then condensate in the syringe, leading to the overestimation of the carbon dioxide concentration.

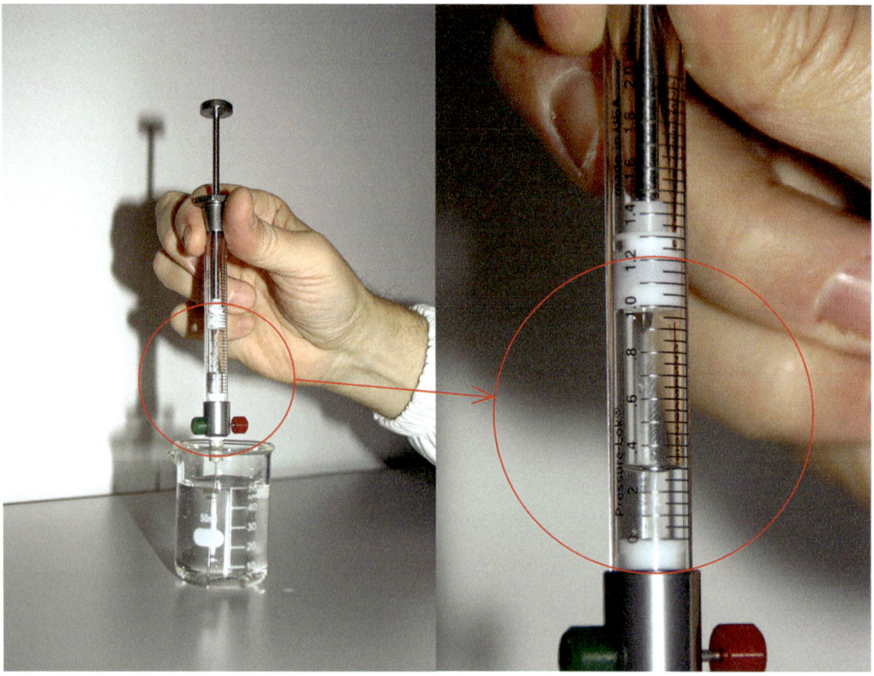

FIGURE 5.7
The syringe with the plunger pushed back to its initial position.

2 On sampling the gas with a clinical 10-mL syringe, discharge the remnant solution into a 5- or 10-mL burette, class A, and use this to measure the liquid's volume. Such a burette has 0.05 mL graduations, with an accuracy of ± 0.025 mL. For example, suppose we sampled 5 mL of gas and subsequently 5 mL of caustic soda solution. After the CO_2 absorption, the remaining volume of solution, measured with the burette, is 2.15 mL. Consequently, the CO_2 percentage is 2.15/ 5.0 × 100 = 43%. The absolute error in the liquid measurement is 0.025 mL, so the absolute error of the CO_2 percentage is 0.025/5.0 = 0.5%. The CO_2 content in the biogas sample is then 43% ± 0.5%.

5.3 Solving the Problem of the Thermostatic Bath's Evaporation

One of the most interesting features of the instruments for biological tests with built-in data loggers, like the AMPTS II and similar, is the possibility to access the data through the Internet. This allows the biogas plant manager to launch the tests and let the instrument do all the work, so it is not necessary to stay all day in the laboratory. The drawback is that one needs to frequently check the level of the water bath to avoid it drying out. In that eventuality, the automatic protection will switch off the resistance, and the reactors will cool, giving wrong biogas production curves.

The evaporation rate is higher when the laboratory's temperature is low (<20°C, for instance, during weekends in winter), when the air in the laboratory is too dry (RH < 60%), or when conducting experiments under thermophilic conditions. Suppose the volume of the water in the incubator drops to such a level that the reactor's temperature does not remain uniform. In that case, the gas production curves will show conspicuous variations in the fermentation rate that the instrument will detect, giving the false impression of unsteady biological activity. When refilling the water bath to compensate for the evaporation, the incubator's temperature will fall abruptly. Such a sudden temperature change reflects in the methane production curves as "steps" on the general sigmoid, as shown in Figure 5.8.

To ensure the unattended water bath operation for long periods without encountering the issues described above, the solution consists in lowering the evaporative rate. The following paragraphs explain some of the possible strategies to adopt.

5.3.1 Seal the Gap between the Reactors and the Plexiglass Cover by Means of O-rings or Rubber Bands

This operation reduces the evaporative rate, but does not stop it, because it is impossible to obtain a perfect seal.

5.3.2 Add Some Very Soluble Salt to the Water

This trick is based on *Raoul's Law*:

$$P_s = X_a \cdot P_0$$

where P_s = vapor pressure of the solution at the reference temperature; X_a = molar concentration of the water in the solution, in %; P_0 = vapor pressure of pure water at the reference temperature

In other words, the vapor pressure of any aqueous solution decreases when the concentration of the solute grows, because the concentration of the solvent (water) decreases.

For instance, when carrying out an experiment at 40°C with the thermostatic bath full of distilled water, its vapor pressure, P_0, will be 7.37 kPa. Suppose that common salt, ClNa, is added to the water. The maximum solubility of ClNa at 40°C is 366.9 g/l. Considering that 1 L of water contains

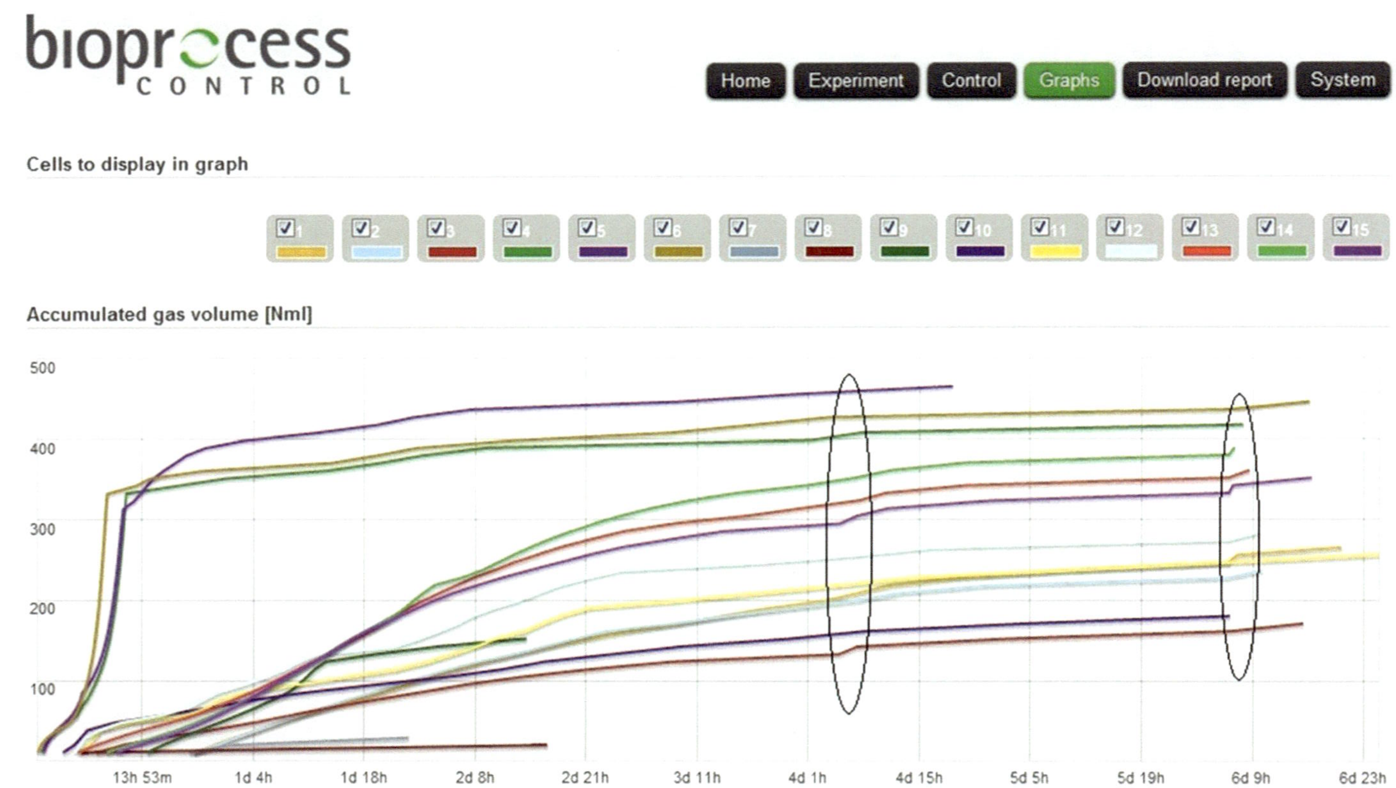

FIGURE 5.8
The steps on the general sigmoid curve are the consequence of temperature variations of the thermostatic bath, caused by the evaporation and subsequent addition of water.

55.55 mol and that 366.9 g of ClNa are equivalent to 6.32 mol, the water concentration in this saturated brine will be:

$$X_a = \frac{55.55}{55.55 + 6.32} = 89.7\%$$

Hence, the evaporation rate of brine will be 89.7% that of distilled water. In other words, the time to evaporate a given quantity of water from the brine will be only 10.3% longer compared to the same volume of distilled water. By employing other salts, for instance, a saturated solution of LiBr (lithium bromide), the increase in time between two recharges of the water bath could reach 20%. The drawbacks of this solution are: triggering corrosion effects in the thermostatic bath when employing NaCl, high cost if employing LiBr or other similar salts, and precipitation of salt in any case. The reduction of the evaporation rate provided by this method is not as high as one should desire; hence, the use of brine is not an optimum solution for the problem.

5.3.3 Replacing the Water in the Thermostatic Bath with Any Fluid Having Low Vapor Pressure

The best commercial candidates for this scope are silicone oil, glycerol (glycerin), ethylene glycol (EG), and propylene glycol (PPG). Silicone oil is expensive and potentially toxic, with the additional drawback of its lubricating quality that would make the reactors difficult to manipulate during the preparation of an experiment. Glycerol has an oily texture too, but its toxicity is low (it is used as a sweetener in the food industry and as a skin moisturizer in the cosmetic industry). Glycerol is biodegradable, and considering the temperature at which the incubator usually operates, there is a risk that aerobic microbes could alter its composition over time, making it unsuitable for our purposes. EG and PPG are both cheap and widely available. EG is available at any petrol station, or big supermarket, or DIY shop, since it is one of the most diffused antifreezing products for car radiators. The toxicities of both EG and PPG are low—slightly better PPG from this point of view—but neither of them is easily degradable by aerobic bacteria. The vapor pressures of PPG and EG at 40°C are 0.045 and 0.039 kPa, respectively. Compared to 7.37 kPa of distilled water at the same temperature, it means that both EG and PPG will evaporate about 200 times more slowly than water.

The thermal capacity of EG is lower than that of water (2.404 kJ/kg · K for EG, compared to 4.181 kJ/kg · K of water), requiring nearly half the energy to reach the temperature setpoint when turning the water bath on. The thermal conductivity of EG is about half that of water, but enough to maintain the uniformity of temperature in the incubator, where the convective thermal exchange is stronger than the conductive one. The low evaporation rate of EG constitutes the main advantage of this liquid on water: the energy consumption of the incubator will be smaller because the latent heat dissipated with the evaporation will be negligible. The author used to consume up to 5 L of distilled water each week in a thermostatic bath working at 38°C. Since he replaced water with EG, the evaporation became almost imperceptible. This represents an energy saving equal to:

$$E_{evap} = 5\,\text{kg} \cdot 2{,}559.1\,\text{kJ/kg week} = 12{,}795.5\,\text{kJ/week} = 14.21\,\text{kWh/month}.$$

Other advantages of EG, that make it preferable to water, are the absence of calcareous scale, unavoidable when employing tap water, and of rust spots, caused by the acidity of demineralized water, since it contains dissolved carbon dioxide.

If you plan to purchase EG at a petrol station or in the car supplies sector of a big supermarket, read carefully the chemical composition on the label, because there are two different types of antifreezing liquids for radiators. One of them is a mixture of EG and anti-corrosion additives (often colored red, but this may change from country to country). The second is EG, usually colored in blue or green, often containing water in different percentages (cheaper products contain more water and must be avoided; search for the product with the highest purity possible, i.e., the lowest freezing point). Both are suitable for our purpose, but do not mix them to avoid undesired chemical reactions between the anti-corrosion and colorant compounds. During the first days of running a test with the thermostatic bath full of radiator fluid, you may observe that the level drops. Add more product until all water has evaporated, and the liquid level in the thermostatic bath will remain stable, just below the plastic lid.

5.4 Checking the Calibration (Volumetric Methods)

We call *calibration* the operation by which a measuring instrument is configured or adjusted to improve its accuracy. The operation requires comparing the measures taken with the instrument under calibration with those taken with a *standard instrument*, or measuring magnitudes of known value, called *calibrators*.

The periodic calibration is necessary for some instruments that tend to lose accuracy with time (e.g., optical or mechanical measuring systems subject to dirt and wear, electronic pressure sensors subject to thermal drift…). In the case of eudiometers and self-built liquid displacement instruments, it is advisable

TABLE 5.1

Density of Distilled Water at Different Temperatures

Temperature (°C)	Density (g/cm³)	Specific Volume (cm³/g)
0	0.9998395	1.00016
4	0.9999720	1.0000280
10	0.9997026	1.000297
15	0.9991026	1.000898
20	0.9982071	1.001796
22	0.9977735	1.00223
25	0.9970479	1.00296

to perform a calibration over the whole measuring range before starting to use them. The extended calibration helps account for the compressibility of the gas because the height of the water column is not always constant, and some of the water contained in the graduated cylinder may evaporate, giving a false reading. Commercial liquid displacement instruments working at constant water column (e.g., AMPTS II and III, BPC Go, MilligasCounter, and similar) are calibrated in their factory and, in general, do not require periodic recalibrations.

When calibrating instruments for measuring the volume of biogas (or of net methane) produced by AD, two possible methods are available: the gravimetric method and the volumetric method.

5.4.1 Gravimetric Calibration Method

In this case, a given water volume injected in an airtight container displaces an equal gas volume (usually air). The said gas volume will be our standard magnitude, so we will then adjust the instrument so that it gives the corresponding reading. The sample volume of gas is measured by *weighing* the quantity of distilled water injected into the container. Hence, the standard instrument is an analytical scale, a kind of instrument usually featuring high precision and accuracy. Remember that 1 mL of distilled water corresponds to 1 g only when the distilled water is at 3.98°C. Consequently, to perform a correct calibration, it is necessary to measure the temperature of the distilled water and to multiply the scale's reading by the corresponding specific volume, presented in Table 5.3. (It is acceptable to interpolate linearly for intermediate temperatures.) Figure 5.9 shows the assembly scheme of a "field" gravimetric calibration system. A good quality moisture analyzer scale, or a scale with at least 1-mg accuracy, could be sufficient as a standard instrument. The recipient with any initial quantity of water should be placed on the weighing pan, and the scale should then be brought to zero. After connecting the tubes (as short as possible) to the instrument under calibration and the water recipient, water can be injected into the latter using a syringe, a burette, or a drop counter.

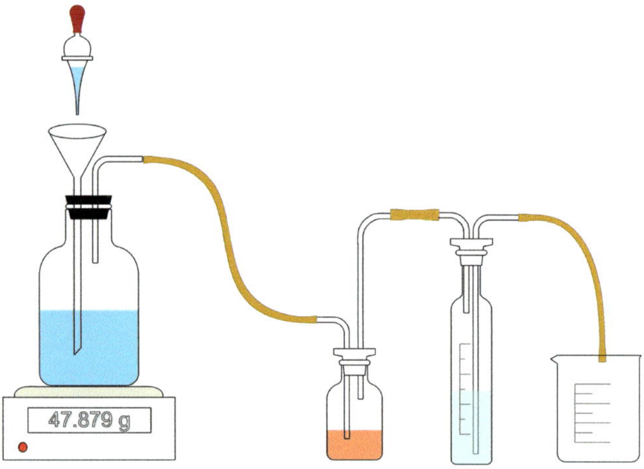

FIGURE 5.9
Scheme of a gravimetric system for routine control of the accuracy of a gas volume measuring device in the field.

5.4.2 Volumetric Calibration Method

In this case, a known volume of distilled water, measured with a burette, displaces an identical volume of gas. The burette is then the standard instrument, less accurate than an analytical scale, but cheaper and straightforward to employ. Figure 5.10 shows the assembly scheme of a volumetric calibration system.

In general, a good calibration of an instrument includes preparing an error table. For instance, when calibrating the liquid displacement system shown in Figure 5.10 employing a 10-mL burette having 0.1-mL divisions, the procedure consists in adding exactly 10 mL each time and reading the volume effectively measured by the instrument under calibration. In the example shown in Figure 5.10, the displaced volume of water is usually measured with a graduated cylinder, having a minimum reading (the space between graduations) equal to 1 mL. We must log the measured values in three columns: the cumulated volume injected with the burette in the first column, the volume measured by the instrument under calibration in the second column, and the relative error of the reading in the third column. Thus, we obtain the distribution of the errors as a function of the reading. The maximum relative error found throughout the whole measurement range will be the value that we must assume as instrumental error for the error propagation analysis. Subsequently, the normalization error must be added to the said maximum error of the measured volume. Eventually, the overestimation error of the reactor's head volume, if filled with inert gas at the beginning of the assay, must be considered too. The normalization relative error is equal to the sum of the relative errors of both the ambient barometer and

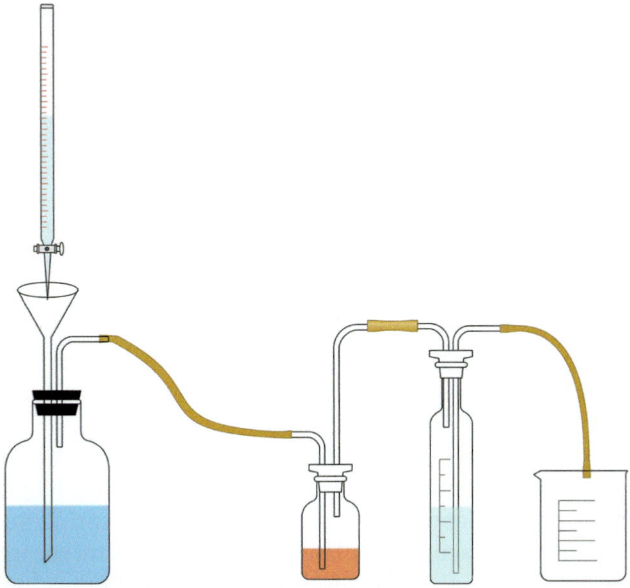

FIGURE 5.10
Assembly scheme of a volumetric calibration system for routine controls in the field.

the thermometer. The overestimation error caused by the moisture content in the measured gas and by the presence of an inert gas is much more difficult to assess. We can estimate it as 5% of the total cumulated gas volume in the usual 30 days, if you are using an instrument that lacks automatic moisture correction (e.g., AMPTS II and AMPST II Light). Modern instruments, such as AMPTS III Standard and Light, have a built-in function for automatically discounting the moisture.

5.5 Flushing the Headspace of the Reactor with a Gas Lighter Recharge

Before launching a batch test, it is a usual procedure to purge the head volume of the reactor with an inert gas, to eliminate oxygen and allow the anaerobic digestion to start immediately. The most commonly employed flush gas is N_2, although some researchers defend using a mixture of 60% N_2 and 40% CO_2. Other inert gases, like He, can be employed too. Purging with pure CO_2 is not advisable when the reactor is connected to a filter with caustic soda, because the latter will quickly absorb the gas, creating a partial vacuum in the headspace. The partial vacuum could suck caustic soda back to the reactor, compromising the result of the test. Another undesirable effect of the vacuum created under said conditions is the saturation of the headspace with water vapor, which may condense in the tubes and dilute the caustic soda solution.

In general, commercial biogas plants do not keep N_2 cylinders in their warehouse. Is it mandatory to purge the headspace with an inert gas? The answer is no. The reason why research laboratories employ N_2 or He is due to safety reasons. In research laboratories, it is common to find several researchers sharing a space where the use of open flames, such as those from Bunsen burners, is possible. Flushing with inert gases eliminates the risk of accidental fires. In a biogas plant, the situation is the opposite—there cannot be open flames; one is usually alone in the lab, and there is plenty of biogas that can be taken from the digester. Suitable containers for sampling enough biogas for flushing the reactors can be: a gas sampling balloon, a "maxi" gas syringe (much more comfortable to use but expensive), or a capsized laboratory bottle within a water basin. Aspiration hoods are not usually present in biogas plants, as is the case in research laboratories. Nevertheless, in general, there is no problem in opening a window or ensuring good ventilation with a small air extractor in a plant's laboratory. Furthermore, the headspace volume of laboratory reactors is minimal, so even if purging them with biogas, methane, or other combustible gases, it is physically impossible to create an explosive atmosphere.

The author usually purges the reactor's headspace with a gas lighter recharge. Lighter recharges usually contain propane, or a mixture of propane/butane, both hydrocarbons similar to methane, perfect for creating an anoxic condition. Lighter recharges are small cans with a safe valve, easy to handle, and available at any tobacconist's for a few dollars. We must remark that the headspace is initially filled with air, which contains just 21% oxygen, the rest being nitrogen. Hence, if we flush the headspace with twice to five times its volume of propane, the proportion of oxygen remaining after the purge will be so low that it will not harm or delay the digestion process. From a safety point of view, if a quantity of propane in the range of 200–500 mL diffuses in the air of the room, there is absolutely no risk of explosion. The *lower explosion limit (LEL)* of propane is 2.1%, so even in the case of a tiny room lacking any ventilation, for instance, 9 m^2 with 2.4 m height, at least 450 L of propane would be necessary for forming an explosive mixture. Considering that lighter recharge cans contain around 42 g of butane, which means 16.5 Sdm^3 (density = 2.5436 kg/Sm^3), one should fully discharge into the room's atmosphere the content of 27 such cans before reaching dangerous levels.

Figure 5.11 shows a gas lighter recharge, available at the tobacconist for a few dollars, being employed to purge a laboratory reactor's head volume.

After flushing the reactor's headspace, but before starting the test, the laboratory operator must check his/her instrument's settings: incubation temperature, stirrer's speed, resetting the gas measure unit, etc. Since their second generation, the AMPTS Standard and Light, Gas Endeavour,

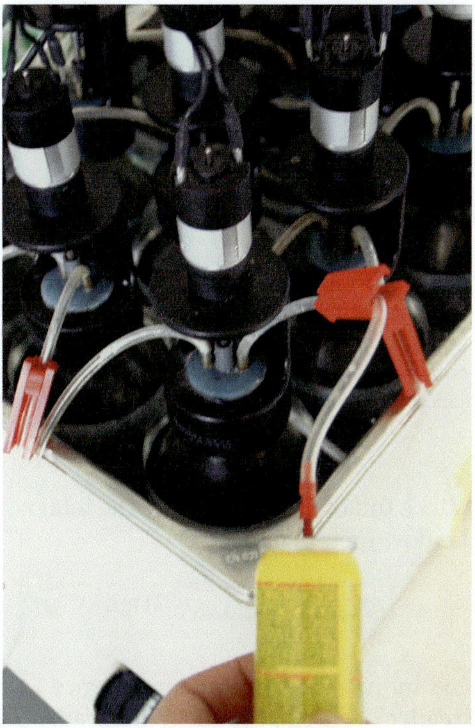

FIGURE 5.11
Flushing the air from a laboratory reactor with a gas lighter recharge.

and Bio Reactor Simulator offer the possibility to activate an algorithm that accounts for the eventual partial vacuum caused by the absorption of carbon dioxide by the caustic soda solution. If employing biogas from the plant to flush the headspace volume of the laboratory reactors, it is necessary to define its CO_2 concentration together with the other instrument's settings. If employing any inert gas or butane from a lighter recharge, then it is required to set 0% in the corresponding instrument's software screen (Figure 5.12).

The simplest option when the reactor's headspace is small (100–200 mL) in comparison with the quantity of methane that the substrate will presumably produce (at least 1500–2000 Ncm^3 in a 500-mL reactor), consists in just not flushing the headspace volume at all. For instance, considering that in 100 mL of headspace, only 21 mL is oxygen (the rest is nitrogen), this oxygen will be converted into an equivalent volume of carbon dioxide by the facultative bacteria. The caustic soda solution will immediately absorb the CO_2 thus produced; hence, the test will start with a light depression and a modest increase in the gas's moisture, which will eventually condensate in the same reactor's headspace or the filter with caustic soda. At the end of the test, all

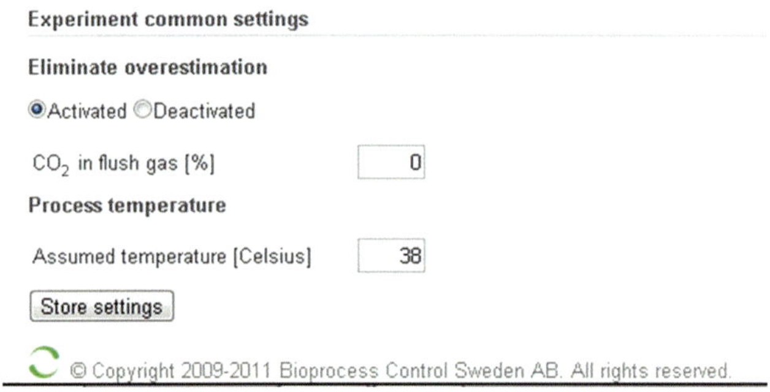

FIGURE 5.12
Setting the initial experimental conditions in an instrument equipped with an algorithm to eliminate the overestimation of the measured gas volume. (Courtesy of BPC Instruments AB.)

the head volume will be full of methane, hence the total volume of gas effectively counted by the flow meter will be:

$$V_{\text{measured}} = V_{\text{produced}} - 21\,\text{mL}$$

The error introduced by not flushing the air is hence 21 mL/1000 mL, i.e., 0.21%. Should the reader decide not to flush the headspace at the beginning of the test, and in the case he/she is employing an instrument with overestimation compensation algorithm, like the already mentioned AMPTS II and similar ones, the option "eliminate overestimation" shown in Figure 5.12 must be turned off.

5.6 Finding Gas Leaks

Gas leaks are a frequent cause of errors. According to the IWA protocol on BMP assay (see Chapter 6), it is advisable to check and identify eventual gas leaks in all reactors before starting any test. It is challenging to locate gas leaks when employing volumetric gas measurement methods because these work at near-ambient pressure, preventing the use of soap bubbles or similar solutions to find the leak points. The following pictures show the procedure step-by-step in the case of a 500-mL standard batch reactor by BPC Instruments. The present method, with any eventual adaptation, is suitable for both barometric and volumetric measurement systems.

First step: Procure a syringe with a volume at least half of the reactor's headspace. Pull the piston up to its end to fill the syringe with air, open

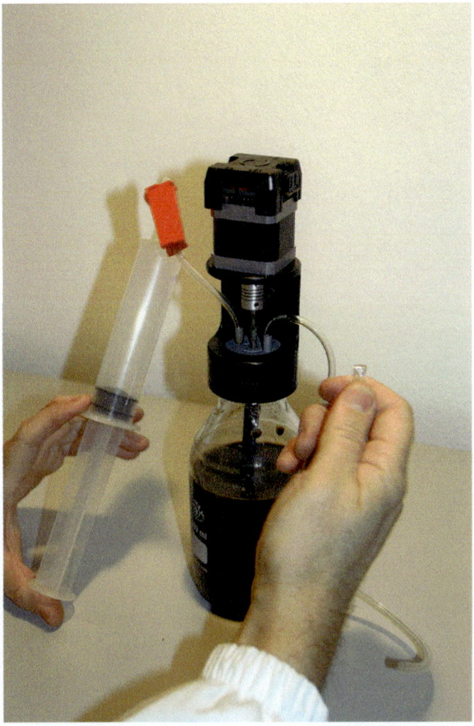

FIGURE 5.13
Setup of the test ready to start.

the flush gas valve in the reactor (red clamp in the photo), and connect the syringe to it (Figure 5.13).

Second step: Choke the tube that connects the reactor with the measurement system, either by hand or with the aid of another laboratory tube clamp. Push the syringe's piston to the end—or as firmly as you can—and hold it for at least 60 s (Figure 5.14).

Third step: Release the piston and wait for a few seconds. It should slowly return as the air in the headspace expands back to its original pressure (Figure 5.15). If it does not return, then there is a gas leak. Check or replace the reactor's cap, the connection tubes, or the clamp(s) and repeat the test. N.B. It is unlikely that the syringe's piston returns exactly to its initial position because of the friction between the syringe's walls and the piston's rubber head. Please observe the difference in the piston positions in Figures 5.25 and 5.27.

Fourth step: Repeat the same procedure with the CO_2 traps (if you are employing a system equipped with them) and eventually with your gas measurement device.

FIGURE 5.14
Pushing the syringe's piston to the end to pressurize the reactor's headspace. Ensure that the tube clamp in the flush gas port of the reactor (red plastic clamp in the photo) is open, and that the exit tube is correctly choked (eventually place an additional clamp on it and close it tightly).

Hints:

1 When checking barometric measurement sets, it is enough to pressurize the reactor, write down the initial pressure, and check it after some minutes. If the pressure decreases, then repressurize the reactor and apply a soapy solution with a small paintbrush to all part joints. Small soap bubbles will show the points from where the gas leaks.

2 If the volume of the available syringe is much smaller than the reactor's headspace, then one piston stroke will not raise the pressure enough to overcome the friction when the piston is released, giving the false impression that there is a gas leak in the reactor. In that case, it will be necessary to pressurize the headspace in steps. Aspirate air, then connect the syringe to the valve, open the valve, push the piston to the end, close the valve, remove the syringe, aspirate air, and repeat the cycle as many times as necessary to have enough pressure in the headspace.

FIGURE 5.15
After releasing the syringe's piston, it will slowly return because the gas initially compressed in the reactor's headspace acts like a spring. In this example, the friction caused the piston to return to about 70 mL instead of the initial 100 mL. This is normal because of the friction losses. Had there been a leak, the piston would not have returned when released.

5.7 Some Safety Rules for the Biological Laboratory in the Biogas Plant

Performing the simple biological and physicochemical tests described in this book does not create any particular risk beyond those usually encountered during ordinary household work. The following safety advice is not a prescription from any norm or regulation, but just common sense.

1 Keep domestic animals and children away from the laboratory.
2 Always wear disposable latex gloves and eye protection. This is mandatory when handling caustic soda, acids, or sodium hypochlorite (bleach).
3 Caustic soda solutions should always be prepared with distilled or deionized water, adding the soda gradually to the water and stirring

all the time. Avoid using tap water, since it will induce some precipitation (mostly carbonates). Adding too quickly caustic soda to water will most likely produce fumes from it. The same contain aerosolized sodium hydroxide, because the dissolution is exothermic and may give rise to microbubbles of water vapor. Avoid breathing such fumes, since they are highly irritating.

4 Always work in a well-ventilated room. If an aspirating hood or an air extractor is not available, at least open a window or a door.

5 When working with the muffle oven, the crucibles or other objects will be at 550°C, which is more than double the usual temperature of a kitchen oven. Always handle the hot objects with a special laboratory tong, whose length must be such that the operator's hands remain outside of the muffle's cavity (Figure 5.28).

6 Do not smoke or eat in the laboratory.

7 Do not fill beverage bottles with potentially harmful liquids like caustic soda solutions or acids: intoxication and burning risk in case of a mistake.

8 Abundant fresh, clean water, or a washing basin with an easy-to-open tap, must always be at hand for emergency washing in case of accidental contact with caustic soda or acids.

9 Avoid laying cable extensions on the floor: risk of stumbling upon them.

10 Always keep your work area clean and tidy.

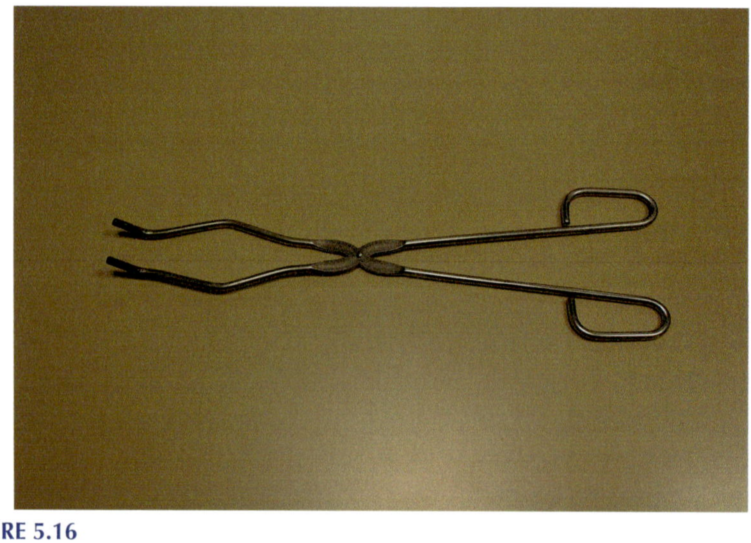

FIGURE 5.16
Special tongs for handling crucibles and other hot objects from the muffle oven.

11 A small blackboard and a piece of chalk often prove handy for writing down an instrument's reading or a memo, especially when one is wearing gloves or has dirty hands.

12 Turn off or mute your cellphone or radio while preparing a test. Distraction sources usually lead to human errors (e.g., forgetting to tare the scale before weighing, exchanging two reactors, etc.).

13 Do not flush the digestate resulting from the tests in the water closet (WC). Although some digestate will not harm the sewage system, it is more sustainable to use it as fertilizer for the garden or to dump it in the same plant's premix tank.

14 The exhausted NaOH solution of the CO_2 filters usually contains some active soda, so it can be employed to wash and disinfect the reactors and tools—except aluminum objects—and then it can be flushed in the WC (it is like the products sold in the supermarket for unblocking drain pipes). It can also be dumped in the premix tank, since it will just add some alkalinity to the sludge.

15 If using NaOCl (bleach) to wash and disinfect reactors and tools (not mandatory, but some people prefer to), follow these simple rules: leave the objects overnight in a solution of one teaspoon (5 mL) of NaOCl per liter of water. After use, the solution can be disposed of in the WC.

16 A dishwashing machine, even an old one, turns very handy in the lab to wash glassware, spatulas, and similar tools. Bottles require hand washing with a brush.

5.8 When and How to Connect the Unidirectional Flow Valves to a Flow Cell

Third-generation instruments, such as AMPTS III Standard and Light, BPC Go, and BRS III, feature interchangeable flow cell units (FCU). Figure 5.17 illustrates a 9-mL FCU mounted on the BPC Go base. FCUs are calibrated and sealed in the factory. The cover features two gas connectors, one for input and one for output, with a 4 mm outer diameter. The diameter is designed to allow the connection to the reactor using gas-tight Tygon (units sold until 2023) or polyurethane tubes (e.g., FESTO PUN-H-6X1-NT, current solution). Such tiny connectors are relatively delicate and can be damaged when connecting or disconnecting the tube at the beginning or end of a test. To avoid accidental damage, the best solution is to leave the cell permanently attached to a 10 cm piece of tube. The tube will be equipped at the opposite end with a fast push-in connector (e.g., FESTO QS-4, as shown in Figure 5.17), allowing you to connect the FCU to either the carbon dioxide absorption flask or the reactor,

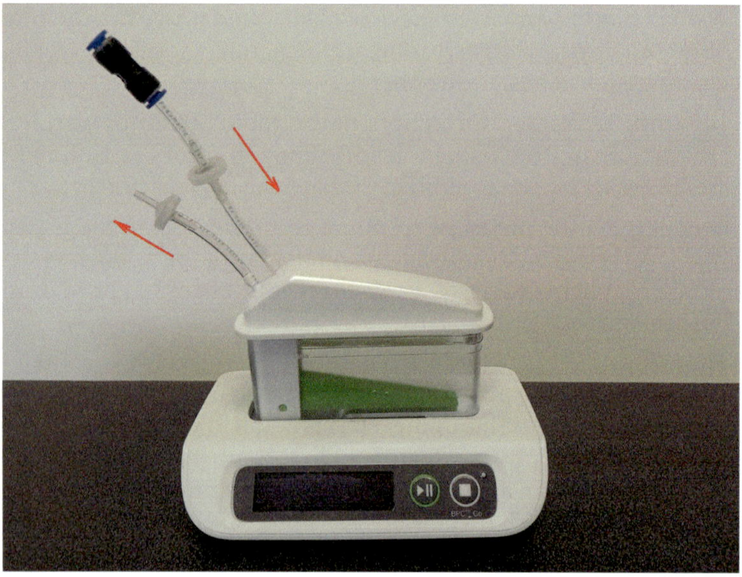

FIGURE 5.17
The correct assembly of the unidirectional flow valves in instruments with sealed flow cell units.

depending on the test being carried out. When conducting batch tests, the atmospheric pressure may become slightly higher than the pressure inside the reactor. This occurs when the production of CO_2 exceeds the production of CH_4. The flask containing caustic soda will readily absorb CO_2, and the partial pressure inside the reactor's headspace will decrease, aspirating water from the FCU. Water may eventually reach the CO_2 absorption flask and dilute the soda. If the reactor's pressure rises and/or the atmospheric pressure drops, caustic soda solution may flow to the FCU and damage it. When performing batch tests, it is mandatory to insert two unidirectional flow valves as shown in Figure 5.17 to prevent such a situation. The valve is shaped as a flat cone, with the vertex pointing in the flow direction.

When performing continuous tests, CO_2 absorption flasks are seldom used, and the gas flow is expected to be constant; therefore, connecting the check valves to the FCU is still recommended but not mandatory.

Bibliography

Japan Soda Industry Association, *Safe Handling of Caustic Soda*, November 2006.
MEGlobal, *Ethylene Glycol Product Guide*, 2013.

Nistor, M., *Calibration Protocol for the Gas Volume Measuring Device from AMPTS II*, Manual for internal use of Bioprocess Control Sweden AB, Version 1.1, August 2011.

Rosato, M.A., *Optimising Biogas Plants – Squeezing profit Out of Your Digester.* CRC Press, UK, 2025. www.routledge.com/Optimising-Biogas-Plants-Squeezing-Profit-Out-of-Your-Digester/Rosato/p/book/9781041003298

6

Critical Review of the Scientific Literature from the Biogas Plant Manager's Perspective

6.1 Introduction

The goal of the present chapter is to assist both biogas plant managers and academic researchers. The former will find a valuable method to determine whether information and data found in scientific journals can be applied to their plants. In contrast, the latter will find some examples of conceptual errors or misconceptions that should be avoided or at least considered before publishing research results in peer-reviewed journals.

6.2 Three Methods to Find Out Absolute Truths: The Aristotelian Syllogism, the Cartesian Doubt Principle, and Avoiding Logic Fallacies

Biogas plant managers, especially when the plant's main commercial scope is energy production, are usually under pressure to increase profit. This implies avoiding any modification to the routine that can potentially upset the process and lose profitability. On the other hand, there is considerable pressure to find new ways of producing more methane per ton of feedstock and/or per cubic meter of plant capacity. It is human nature to be afraid of the unknown, but making things always the same way brings no improvement. We have seen in the previous chapters how to apply Lord Kelvin's philosophy—"To measure is to know, and if you don't know something, you can't improve it"—to the biogas plant management, with the help of some simple, yet scientifically sound techniques. Since the Internet offers tons of research papers to download with a few clicks, often free or at an affordable cost, the temptation to save labor and time by just implementing "discoveries" made by universities and other research bodies is high. However, applying published

DOI: 10.1201/9781003676201-6

research data to a biogas plant, without first checking in the laboratory the suitability of such data, has a high risk of causing some damage to the anaerobic process for the following reasons:

1 There is no guarantee that the inoculum and test conditions employed for the published research are comparable to those of the biogas plant; hence, trying to optimize a biogas plant based on a few published tests can yield the opposite result and cause productivity and profit losses.
2 Error analysis seems to be a kind of taboo for the academic world researching biogas, despite the consistent and well-proven set of ISO norms on error analysis and correct expression of the uncertainty of measured magnitudes. Quite often—practically always in the literature about anaerobic digestion—scientific papers do not include the analysis of *measurement uncertainty* but just the *dispersion* of a given number of replicate tests. Sometimes papers do not even contain a thorough description of the instruments employed, and in some cases, the resulting gas volumes are not even normalized. Therefore, one cannot be certain whether the "improvement" reported by the researcher is genuine or merely a result of measurement errors that led to a measured value higher than previously published ones.
3 Since the job of a researcher consists of creating a simplified model of a complex reality, sometimes the simplification is too much and the resulting model is valid only under very restrictive conditions.
4 Research institutions, even public ones, need money to carry on their activities. Often, sponsor companies fund research laboratories to develop or improve a specific product. While such an approach is legitimate, there is a strong probability that marketing claims present the research results in a biased or even distorted manner. The most common marketing hype is generalizing the results obtained with just one particular case to all biogas plants.
5 Quite often, the "test results" presented as proof of the efficacy of a given product or method are based on the *post hoc* logical fallacy that will be explained in detail in the following paragraphs.

Adopting any new method, technique, or feedstock in one's biogas plant requires a preliminary selection process based on the following analysis method:

1 *Reputation of the source*: It is clear to everybody that a paper published by a university or research institute is more likely to be scientifically correct than information published in a blog, even though the publication by a reputed source is not an absolute guarantee of truth or correctness, or applicability to a particular case.

2 *Applying correctly the Aristotelian syllogism*: Syllogism (from ancient Greek συλλογισμός, syllogismòs, compound word of σύν, syn, "together, joint, linked", and λογισμός, logismòs, "calculation, reasoning": hence "linked reasoning") is the base of logic and hence of the scientific method. Given two assumptions, any conclusion derived from these *can be true* only if both are true. The difference with digital or Boolean logics is that in real life, there is no guarantee that a conclusion derived from two true suppositions is *always* true. In logic terms, deriving a conclusion from two true hypotheses is a necessary condition but not sufficient. On the other hand, a syllogism based on a false assumption may lead to an apparently true conclusion, but just by pure chance.

3 *Applying the Cartesian doubt*: Also known as methodological skepticism, this method consists of four steps:
   a Accepting only information you know to be true;
   b Breaking down these truths into smaller units;
   c Solving the simple problems first;
   d Making complete lists of further problems.

4 *Discarding results based on causal fallacies*: The correct way to measure anything is to compare the unknown magnitude to a standard one or to check the experiment's hypothesis against a control experiment run simultaneously under controlled conditions. Comparing the energy output of a biogas plant in a given period with the energy output of the same plant over another time period is the most common causal fallacy in the literature.

The following practical examples will show how to apply the methods described earlier to decide if an academic study is worth being implemented in one's biogas plant.

6.2.1 The Correlation between BMP and Electrical Conductivity

This method was introduced in Chapter 1, Section 1.4.7., and in the present section, we will show how to check its validity rationally.

The correlation between electrical conductivity and performance of the biogas plant has been introduced by a group of Italian researchers (Garuti et al., 2014) as a possible method to "easily control" an anaerobic process. The reasoning presented by said authors is somehow reductive, although formally logical:

1 The conductivity is proportional to the concentration of dissolved salts and to the concentration of ammonium ions, too (true).

2 Both concentrations have a negative influence on the metabolism of the methanogenic archaea and also on the metabolism of other microorganisms involved in the digestion process (assumed true).

3 Hence, it would be possible to establish a correlation between the electrical conductivity and the methane yield of the whole process, tabulate (or plot) it, and employ such a table (or curve) as a control parameter for the plant.

Since its publication in a specialized agricultural magazine, many Italian biogas plant managers have adopted the said method as an absolute rule, blindly measuring the electrical conductivity and logging the measures in historical series. The reason for such acceptance is the unquestionable reputation of the research institute in the Italian biogas sector, the apparent correctness of the syllogistic reasoning (two true statements leading to a conclusion), and a set of experimental values confirming the conclusion. Nevertheless, the authors state in the same article that some tests performed in real-sized plants did not yield such a linear correlation between electrical conductivity and biogas yield.

Applying the methodological skepticism leads to the following reasoning and points out what experimental evidence to search for the following:

1 *Accepting only information you know to be true*: One could argue that archaea are organisms that have a marine origin, and hence they should be salt-tolerant. The author has personally tested sludge from the marine bottom that had good methanogenic activity. Another argument that gives rise to a doubt is that "salt" does not necessarily mean "chloride". The Italian researchers used a "reference saline solution" containing chloride salts of sodium, potassium, and magnesium to create inocula samples with different conductivities. It is known that different salts (more specifically, different ions) have different degrees of toxicity to microorganisms. Furthermore, there are thousands of microbial species thriving together in the digester's sludge, each one having different tolerance levels to salinity. Finally, a given level of electrical conductivity may not cause the same toxicity for the bacteria, depending on whether the cause of said conductivity is dissolved salts, ammonia, or other salts different from chlorides. In a "normal" biogas plant's sludge, the conductivity is given by ammonia, carbonates, phosphates, and sulfates rather than chlorides. The researchers adopted such "reference saline solution" because it is specified in the norm UNI EN ISO 11734:2004 *Water quality—Evaluation of the "ultimate" anaerobic biodegradability of organic compounds in digested sludge—Method by measurement of the biogas production*. The norm applies to wastewater treatment plants, where chloride salts are more likely to be present than in agronomical biogas plants. Consequently, assuming that the normalized method for testing sewage sludge will yield the same result when applied to agricultural digestate, which is characterized by a different microbial population, is a logical fallacy.

2 *Breaking down these truths into smaller units*: Then the proposition (2) of the syllogism at the root of the electrical conductivity method may not be an absolute truth, and if so, the conclusion derived from it cannot be an absolute truth, i.e., it can be true sometimes but sometimes not. Since the "model" of salinity tested by the researchers consisted of adding the aforementioned saline solution until the desired electrical conductivity in each set of test reactors was reached, it is necessary to prove that the same conductivity will produce the same inhibiting effect even when the salts employed are not chlorides. So the doubt is as follows: Was the electrical conductivity the real cause of the impaired methanogenic digestion observed by the researchers, or was it just the observable effect of the chloride concentration, the toxicity being caused by the latter? Finally, like all papers in the biogas literature, the article in question does not present an analysis of the uncertainty, so one could argue that the inhibition observed may be—at least in part—just the result of instrumental or method errors.

3 *Solving the simple problems first*: The first problem to solve is whether a reactor, doped with phosphates and carbonates up to a given electrical conductivity, will show the same decrease of methanogenic potential encountered by the researchers in question. If it is so, then the proposition (2) of the syllogism is more likely to be a "universal truth", and the method of the electrical conductivity monitoring would be "universally" valid. The author then prepared a comparative test with two sets of reactors. The first set consisted of a blank reactor, a reactor with acetic acid, and another reactor with cellulose, all of them inoculated with "natural" sludge having 9.8 mS/cm and pH = 7.6. The second set of reactors was loaded with the same inoculum, to which phosphoric acid, magnesium carbonate, and potassium hydroxide were added until reaching 20 mS/cm, while keeping the pH at 7.6. At such conductivity, the expected BMPs of the second set of reactors should be 15% lower than the control set. Since the author's instrument is an AMPTS II with 1% error margin, such a difference between the reactor sets should be easily and reliably detectable. The results are shown in Figure 6.1.

4 *Making complete lists of further problems*: According to one of the many variants of Murphy's Law, trying to solve small problems leads to bigger problems. Although such a statement has no scientifically proven validity, it seems to be as inexorable as any law of physics. Figure 6.1 shows that there was indeed a reduction of the *calculated* BMP in the case of reactors with higher electrical conductivity, which seems to correspond well with the experience of the Italian research group. But we should consider all the facts affecting our experiments, and then carefully analyze Figure 6.2. It is evident that the blank reactors with the high-salinity inoculum produced more methane

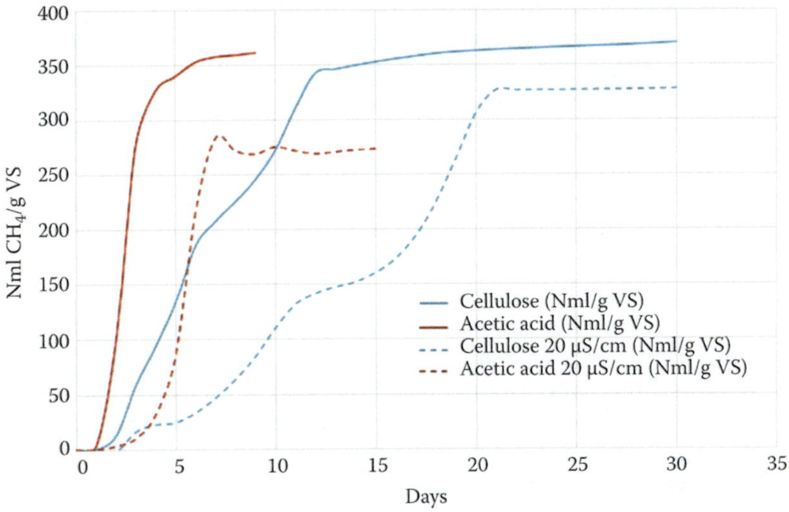

FIGURE 6.1
Specific methane production of the comparative test between two samples of sludge with different electrical conductivities and the same reference substrates.

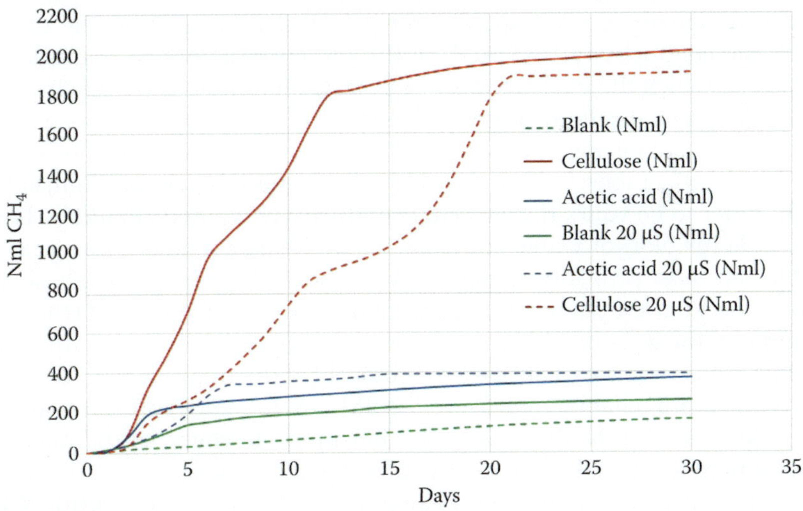

FIGURE 6.2
Gross methane production of the comparative test between two samples of sludge with different electrical conductivities and the same reference substrates.

than the corresponding blank with the "natural" inoculum, and the same happened to the reactors with acetic acid. Such a result is exactly the opposite of what was expected. The shapes of the curves clearly show that the abrupt change in salinity induced by the addition of phosphoric acid and potassium hydroxide leads to some kind of biological disruption, followed by the adaptation of the bacteria to the new conditions. The "raw" curves of the reactors loaded with acetic acid and those with cellulose end up with practically the same total methane production from the same amount of substrate, albeit with different kinetics. The higher salinity induced by P and K seems to have increased the activity of the methanogenic archaea, and hence of the inoculum contained in the blank, and somehow reduced the activity of the hydrolytic bacteria, as we can deduce from the lower production of the reactor with cellulose. Hence, the test led to an uncertain result because the differences in the *calculated* BMP were caused by a higher methane production of the *blank* and not by an impaired methane production of the *sample reactors*. There are many reasons why this may have happened. Still, from the biogas plant manager's perspective, our test shows that the electrical conductivity criterion is not as linear as assumed by the Italian researchers, and its practical application is not straightforward.

6.2.1.1 Conclusion

The experimental results of anaerobic digestion (AD) at 20 mS/cm electrical conductivity do not show a direct correlation between the case in which the conductivity is caused by the presence of sodium chloride and the case in which the ions are those of potassium and phosphorus. In a typical biogas plant, it is quite improbable that sodium chloride reaches relevant concentrations. In general, eventual increases in conductivity will be caused by a "cocktail" of ions: sodium, potassium, magnesium, ammonium, sulfur, etc. In order to be valid, the electrical conductivity method should be based on a table (or plot) of values measured with the actual mix of ions that the sludge can most probably contain. Furthermore, the measurements should be performed at the same operational temperature of the plant. Hence, one should employ a set of tables (or or a more complex algorithm) rather than a single table or curve to apply the electrical conductivity method correctly. It is much more reliable to measure the BMP of reference substrates at periodic intervals with the digester's own inoculum, to check its *real* microbial activity, than to try to deduce it from an indirect measure.

A final reflection for readers belonging to the academic world: When researching the correlation between the salinity of the sludge and its microbial activity, the mineral content of the feedstock should be taken into account. For instance, biogas plants fed with corn silage (like most European ones), equipped with sludge recirculation, will be more prone to the accumulation of salts with time than plants without sludge recirculation, in which the dilution of the feedstock is mainly done with water. Proper research on the correlation between the electrical conductivity and the methanogenic activity of the sludge should be performed using a saline solution, which is representative

TABLE 6.1

Mineral Content of Corn According to FAO (Average of Five Samples)

Mineral	Concentration (mg/100 g DM)
P	299.6 ± 57.8
K	324.8 ± 33.9
Ca	48.3 ± 12.3
Mg	107.9 ± 9.4
Na	59.2 ± 4.1
Fe	4.8 ± 1.9
Cu	1.3 ± 0.2
Mn	1.0 ± 0.2
Zn	4.6 ± 1.2

of the kind of salts found in the sludge itself. For instance, according to a study by the Food and Agriculture Organization (FAO), the mineral fractions contained in corn biomass are those shown in Table 6.1. Hence, the saline solution added to the sludge to increase its electrical conductivity should contain the same minerals, and in the same proportions as listed in the said table.

The salts should be mainly carbonates, phosphates, and sulfates, which are the most likely species found in agricultural sludge. Sodium should be present both as bicarbonate and chloride. A more pragmatic approach could be calcining a certain quantity of biomass in the muffle oven and then using the ash to prepare the saline solution for the tests. Such a saline solution will surely be more representative of the specific digester's operational conditions than the standard saline solution defined by norm UNI EN ISO 11734:2004 that was adopted for the basic research quoted at the beginning of this section.

6.2.2 Assessing the Validity of the VFA/TA (FOS/TAC) Method

This method was introduced in Chapter 1, Section 1.6. Its wide diffusion can be attributed to the marketing efforts of the company that produces the most popular *ad hoc* automatic titrator and the sales argument, "Thousands of biogas plants in Germany use it". The limited validity of the said method is recognized in the German norm VDI4630:2016, page 7:

> The FOS/TAC is substrate- and plant-specific and therefore should not be used directly to compare plants.

The scarce validity of the FOS/TAC method can be proved by applying the simple rules of the syllogism:

1 The volatile fatty acid (VFA) concentration is an indicator of the stability of the AD process (true, but only under the condition that the

concentration of propionic acid is negligible compared to that of acetic acid).

2 The alkalinity is an indicator of the resistance to changes in pH (true).

3 Hence, by correlating the VFA concentration and the alkalinity, for instance, through a simple ratio, we can obtain a measure of the stability margin of the AD process [only when (1) is true, but under certain conditions it can be a misleading conclusion].

The experimental evidence already presented in Section 1.6 should be enough to prove the Cartesian doubt: The FOS/TAC (*Flüchtige Organische Säure/Totales Anorganisches Carbonat)* cannot be universally applied because it is not based on universally valid suppositions. From a practical point of view, a method that tells us that everything is running properly when we already know that there is no problem—but in many cases does not provide any helpful information to diagnose why the anaerobic process is blocked—is unreliable. By applying methodological skepticism, the reader could argue that the case study presented in Section 1.6 is just one example of a biogas plant that, for some reason, suffered a severe inhibition of the archaea. Hence, the following experiment will dissipate any doubt:

1 Take some digestate from a "healthy" biogas plant and sieve it with a 5-mm mesh (a kitchen sieve like the one employed for spaghetti will do).

2 Fill two reactors with an equal amount of digestate and incubate them for at least 5 days.

3 Add 2 g of acetic acid per liter of inoculum to one of them.

4 Add 2 g of propionic acid to the other.

5 Measure the VFA/TA (total alkalinity) of both if you have any doubt. They should be approximately the same, since the VFA/TA method does not make any difference between different VFA.

6 Incubate the reactors for at least 4 days and check how much methane each one has produced.

7 Convinced now?

6.2.3 The Causal Fallacy

A logical fallacy is a false reasoning that seems to make sense. The most common type is the causal fallacy, in which the cause of an observable effect is incorrectly identified. Classical logics call it the *post hoc* fallacy because the Latin sentence describing it as *post hoc, ergo propter hoc* ("after this, therefore because of this"). This is the kind of sales argument usually employed by producers of "boosters", additives, or machinery that allegedly increase the efficiency of an anaerobic process. From a scientific point of view, the differences observed in the yield of a biogas plant before and after applying a specific product can have tens of causes, which are not related directly to the single cause assumed as the source of the differences observed before and after. The following example shows how to correctly evaluate the efficacy of enzymes, promoted by some articles on the basis of "before" and "after" tests in full-sized biogas plants.

6.2.3.1 *Information from the Literature*

1 According to M. Plöchl et al., applying enzymes reduces the viscosity of the sludge, though the effect is negligible. The paper does not include any consideration of the biogas yield, only the advantages of reduced viscosity, i.e., a lower self-consumption of energy in the stirring system.

2 According to Bruni and Agelidaki, treating manure and straw with calcium oxide (CaO) increased the biogas yield by 66%, while steam treatment followed by enzymatic treatment only gave 35% increase.

3 According to the chemical company Dupont, applying their enzymes increases the biogas production by 15%; but in another paragraph of the same brochure, it is stated that the net methane production increases only 8%.

4 According to Petta et al., the addition of enzymes to a biogas plant enabled saving 8% of the feedstock for the same energy production.

6.2.3.2 *Checking the Logic Flaws*

Papers (1) and (2) contain research performed under laboratory conditions, i.e., comparing control samples with enzyme-treated samples, though no estimation of the uncertainty is included. In spite of being scientifically sound, at least from the formal method, the study of Bruni and Angelidaki has little applicability to real-world plant operations. Since they tested the enzymatic treatment after a steam treatment, it is not clear if the increase in biogas production is caused mainly by the first or by the second. On the other hand, publications (3) and (4) are not scientific papers published in peer-reviewed journals and contain "field results" measured directly on real-size plants, "before" and "after", i.e., the method employed raises the suspicion of causal fallacy. In both cases, the enzymes employed are extracted from fungi, i.e., they are mixtures of cellulase and hemicellulase, probably also some laccase. In both cases, the reported improvement is the average of a high number of individual measurements over time. Incidentally, there is no description of the instruments employed for taking said measurements, the measurement method, and the resulting uncertainty range.

6.2.3.3 Checking the Efficacy of Enzymes in the Correct Way

The author performed a simple laboratory test by simply taking sludge from the primary digester of a biogas plant, which contained 9% volatile solids (VS) and performing a comparison between the methane yield of the control reactor and the reactors to which enzymes were added. The sludge contained mainly vegetal biomass and cow manure, hence two different commercial enzymes were tested: one is cellulase in liquid form (300 mg/L recommended dosage) and the other is a mixture of cellulase and hemicellulase (recommended dose 30 mg/L). Figure 6.3 shows the result after 30 days digestion at 38°C. Observe that the curves of the methane production with enzyme addition are almost coincident, showing that both commercial products are functionally equivalent. The increase in net methane yield was 9%. Since the error margin of the instrument is 1%, we can conclude that the application of such enzymes leads to improvements in the 8%–10% range.

An important consideration, not specified in any of the cited publications, is that the activity of enzymes depends on their concentration. This means that the dosage must be calculated in two steps: the initial amount for the first time, as a function of the net available digestion volume, and the daily dose, which depends on the volume of feedstock and water added each day, necessary to keep such concentration constant in time. Example: Suppose a digester has 3,000 m^3 of net digestion volume. The same is fed with 100 m^3 (solid feedstock plus dilution water) every day. The enzyme dosage recommended by the manufacturer is 30 mg/L = 30 g/m^3. Hence, if the plant manager decides

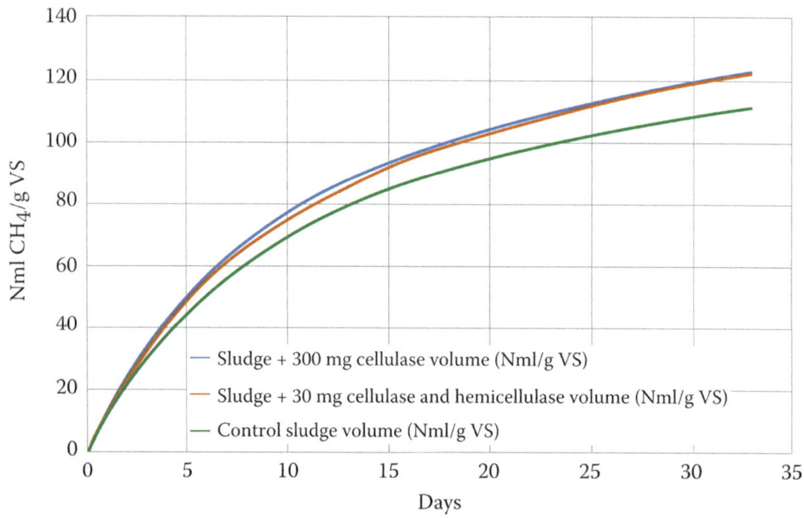

FIGURE 6.3
Comparative AD test of sludge containing high amounts of vegetal fiber, with and without enzyme addition.

to employ such an enzyme, the operation strategy must consider an initial load on the first day equal to 90 kg—in order to bring the whole digester to have an adequate enzyme concentration—and 3 kg per day, to keep such concentration constant.

In this particular case, the results measured in laboratory conditions are in line with those deduced from the published "before" and "after" experiments. Nevertheless, this does not mean that the *post hoc* test method in the plant is equivalent to a test under controlled conditions in the lab. A sales agent once confessed to the author that demonstration tests offered to potential customers led to no tangible results in 50% of the cases. This casts a further doubt about "real-size test": positive results published in specialized magazines, or even peer-reviewed papers, could be tainted with cherry-picking.

6.3 Misconceptions of the Scientific Literature Amplified by the Marketing

6.3.1 The Importance of pH and the Use of Sodium Bicarbonate as a Buffer Agent in Anaerobic Plants

Adding sodium bicarbonate to the digester, with the scope of "keeping a stable FOS/TAC value" or "avoiding acidification", is a relatively common practice in some European countries. Adding sodium bicarbonate serves only to mask eventual biological problems in the digester, without eliminating their causes. Furthermore, the indiscriminate use of sodium bicarbonate, without laboratory tests determining it is effectively needed, represents a superfluous cost and a potential danger for the fertility of the soils that will later receive the digestate.

In principle, adding sodium bicarbonate indeed helps maximizing the production of acetic acid, the primary nutrient of the acetoclastic archaea, but it is also necessary to consider the following matters:

1 Sodium bicarbonate is not the only cheap product for neutralizing the acidity: lime, potash, and magnesium hydroxide are also helpful. The said alkalis are the main constituents of biomass ash.

2 Agricultural biogas plants are often fed, in variable proportions, with three substrates that feature high alkalinity: bovine manure, swine slurry, and chicken (broiler) or hen (layer) dung. In Germany and Italy, instead, a consistent number of plants are still fed exclusively with corn silage, cereal meals, and similar easily fermentable feedstock, with little or no manure at all in the mix, hence having scarce alkalinity. It is then natural that in such contexts, it becomes necessary to artificially provide some alkalinity to compensate for the VFA

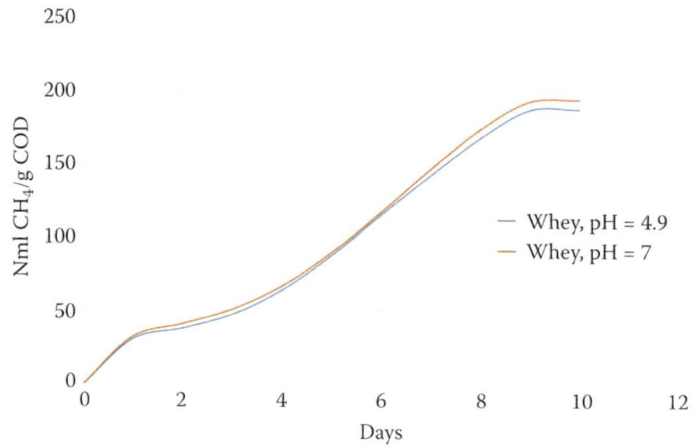

FIGURE 6.4
Comparison between the anaerobic digestion of "raw" cheese whey and the same neutralized with sodium bicarbonate up to pH = 7. (Error margin of this single test = ±5%).

produced by the fermentation. Such a management method should be considered as exceptional, applicable only to biogas plants fed exclusively with corn silage or saccharine or starchy substrates, and not as a rule for all biogas plants.

3 Adding sodium bicarbonate to a digester that shows some tendency to acidity may cause an unwanted phenomenon: foaming. This is just the result of a chemical reaction and has nothing to do with the biological activity.

Figure 6.4 shows experimental evidence of the former considerations. The author measured the BMP of a very acidic substrate, cheese whey. The inoculum was sampled from a "typical" agricultural biogas plant, usually fed with 10% bovine manure and 90% mixed cereal silage, all the said mixture being diluted with swine slurry. A reactor was fed with pure whey, pH = 4.9, and another reactor was fed with the same whey that had been previously neutralized with sodium bicarbonate (pH = 7). We can observe that the difference between the two curves is irrelevant, lying within the typical error margin of the BMP assay.

If the usefulness of sodium bicarbonate is so limited, why is it then so widespread in the biogas industry? Most likely, this is because the technical literature distributed by one of the multinationals promoting it claims the efficacy of its product based on "studies carried out at the University of Rostock". The same technical sheet does not quote the title, nor the author, nor the journal where such studies were published. One of the said studies, published in German with only the abstract in English, can be downloaded from the Internet (see Bibliography at the end of this chapter). From a careful reading, the following aspects appear questionable:

1. The methodology adopted is not rigorous because the tests were performed directly in two 550 kW biogas plants, not in a laboratory. Hence, the published results could be the outcome of mere chance or, worse, cherry-picking.
2. The merit parameter adopted to judge the validity of sodium bicarbonate as an additive is the FOS/TAC ratio (furthermore, not even measured by the researchers themselves, but by the plant's manager).
3. As usual, the paper contains no analysis of the measurement errors, which are presumably high because the measures were taken with instruments homologated for industrial scopes.
4. The text states that the systematic addition of sodium bicarbonate increases the production of methane, but the tables with the results show only the gross biogas production of each period, while the net production of methane is displayed in a graphic, with a tiny scale, and has not been integrated in each period.
5. The BMP of corn silage is expressed in terms of dry matter and not in terms of VS; hence, it is not possible to check if the digester's organic load (OL) was constant throughout the test period.
6. The difference between the *gross biogas production* of both plants—a control plant and a test plant to which sodium bicarbonate was added—results in the same order of magnitude of the (estimated) measurement error. Furthermore, stating the production of biogas instead of the net methane production results in the following legitimate Cartesian doubt: Was the methane content the same in both plants all the time? The paper fails to explain how to fairly compare two biogas plants based on "gross biogas production", as the methane content is inherently variable over time. More precisely, when adding sodium bicarbonate to the slurry under active fermentation, it will react with the VFA, yielding CO_2 that adds to the biogas flow and lowers the CH_4 concentration. Therefore, assuming a constant 60% methane content is a mistake. Figure 6.5 shows that, after a "shock treatment" with sodium bicarbonate, the overall productivity falls drastically and is comparable to that of the control plants only from the 60th day of the test. After 100 days of the test, the plant to which the bicarbonate was added seems to have performed better than the control. At this point, we may ask ourselves if such a result is the merit of the bicarbonate addition, or is just a variability resulting from the combined uncertainties introduced by differences in the management of the corn silage, measurement errors of the load weighting systems, starts and stops of the generators, and, above all, the error arising from the method adopted by the researchers to estimate gas production. For sure, most agricultural biogas plants in Europe usually do not include gas volume meters. In the case of the paper under analysis, it is not clear how the gas production was measured. From a

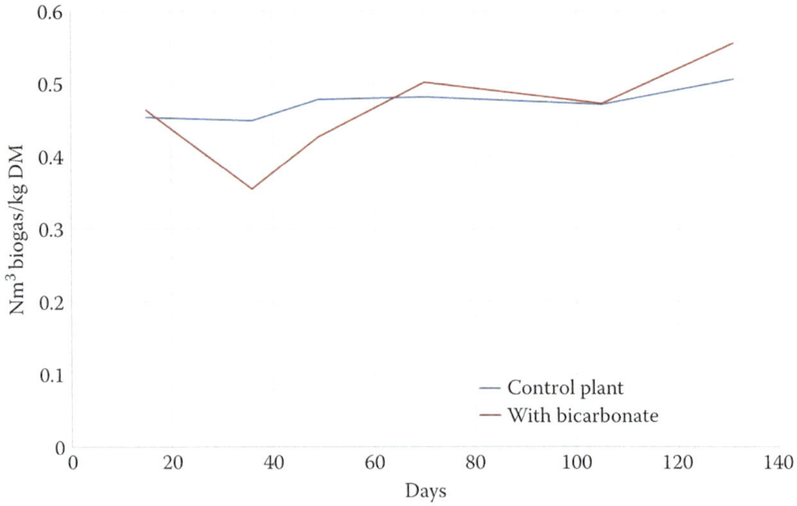

FIGURE 6.5
Gross biogas production measured in two different biogas plants: one had suffered from acidification and was subject to a "shock treatment" with sodium bicarbonate, followed by periodical additions; the control plant received no additive at all. (The graphic was plotted by the author based on the tables published in the quoted paper.)

very synthetic description, it seems that the researchers calculated the biogas flow with a "rule of thumb" based on the inside pressure of the gas collection bag, assuming constant volume. It is evident to anybody, based on just common sense considerations, that flexible biogas holders cannot be considered perfect hemispheres since their volume will vary enormously not only with the inner pressure, but also with the difference of temperature between inside and outside, with the solar radiation, wind pressure, snow load, etc. Should anybody have any doubt: Both the norm VDI 4630 and the Italian norm UNI/T 11703 state that barometric measurement systems must be mounted on *rigid* reactors. Hence, the validity of the conclusions drawn by the authors of said paper is at least dubious.

Another aspect that the biogas plant manager cannot ignore is the accumulation of sodium in the digestate. For instance, the dosage of sodium bicarbonate recommended by the manufacturer for a biogas plant with 1 MW electrical power—in the range of 100–200 kg/day—can in part inhibit the archaea and end up reducing the production of methane. If the dosage mentioned before is kept constant, within 1 year, the total amount of sodium bicarbonate cumulated in the digestate will be in the range of 36.5–73 tons, with the risk of salinizing the soils instead of fertilizing them. Hence, before deciding to systematically add sodium bicarbonate to the digesters, it is highly advisable to ask an independent agronomist to check if the cumulated concentration of sodium can be dangerous for the soil's fertility. The study on soil salinization performed by the FAO and quoted in the Bibliography can provide the reader a guideline for such verifications.

6.3.2 The Hyped Importance of Trace Elements

As in the case of sodium bicarbonate, industries, "biologists", and "gurus" of the AD merrily prescribe the addition of tons of "integrators" and "special products". The sales argument sounds always as an ominous warning: "If you do not add these products, your plant will lose productivity and risks the biological collapse". We have already explained in Chapter 1, Section 1.1.2.6 that it is necessary to supply the archaea with *very small quantities* of some minerals called "micronutrients", "trace elements", "trace minerals", or "oligoelements" depending on the different bibliographic sources. The said small doses of minerals are necessary as catalysts for the enzymatic processes involved in the anaerobic digestion. It is very easy to find on the Internet some "recipes" of mineral mixtures studied by different researchers. Nevertheless, the dosages recommended by the various studies vary enormously (see, for instance, Lebuhn et al., 2008, and compare with Table 8.13, derived from the Italian standard UNI/TS 11703). Not surprisingly, companies selling such mixtures always recommend the highest dosages found in the literature. Some biogas plant managers tend to exceed in the addition of trace minerals, fearing that the productivity of the digester may fall if they do not keep high concentrations of them. The truth is that not all trace elements may be necessary. Furthermore, the highest dosage is not always the best solution. According to Gustavsson (2012) and Yee Yaw Choong et al. (2016), the acetoclastic archaea group requires the availability of trace elements, in particular, Ni, Co, and Fe, while the hydrogenotrophic archaea are inhibited if the concentration of such minerals exceeds a given threshold. It is then clear that the dosage of minerals must respect a very delicate balance—too high will inhibit the hydrogenotrophic archaea, hence the proportion of CO_2 in the biogas will rise. Furthermore, the mere presence of trace elements in the sludge is not enough to grant a good methanogenic activity—the minerals must be present in a form that is soluble in water and that the archaea should be able to assimilate (e.g., sulfates, carbonates, phosphates, etc.). As a further complication, the concentration of H_2S plays a role in the assimilability of trace minerals, since H_2S may react with some metallic salts, precipitating them as insoluble sulfides. The presence of phytic acid, usually in high concentration in those biogas plants fed with cereal silage and very little or no cow manure at all, negatively affects the balance of trace elements. Phytic acid tends to chelate metallic ions, making them unavailable to the archaea. Ruminant's manure contains phytase, an enzyme that breaks the chelates. So, the amount of trace elements that should be added to the digester to

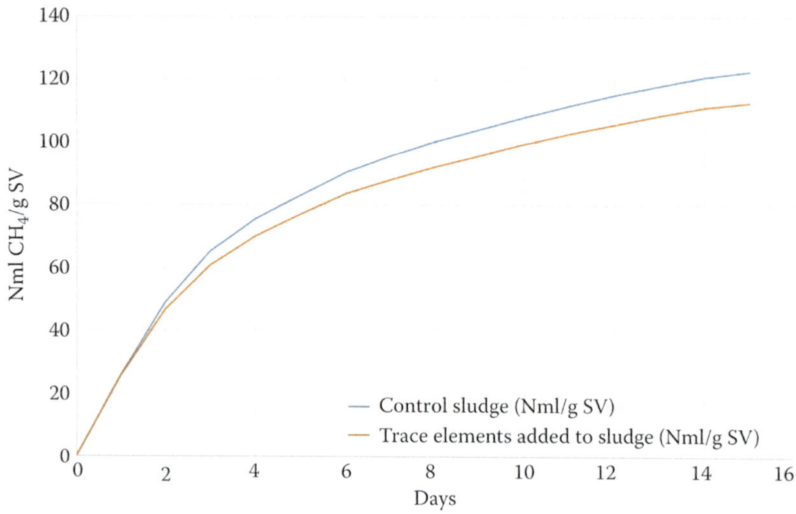

FIGURE 6.6
Effect of adding trace elements to an inoculum that does not need them. The result was 8% less methane in 15 days, compared to the control reactor. The author conducted this test using inoculum sampled from a plant that is typically fed silage, vegetables, olive mill pomace, and cow manure, which showed no signs of biological issues. The test was performed with an AMPTS II, and perfectly preincubated inoculum, resulting in an overall error margin of 1%.

keep its activity at an optimum level is also a function of the percentage of cow manure in the digester's diet.

The easiest way to determine whether a plant needs the addition of trace elements and their optimum dosage is to perform several specific methanogenic activity (SMA) tests on the same inoculum, each one with a different concentration of added minerals. Figure 6.6 is an example of such a test. In this particular example, the inoculum already had a sufficient concentration of trace elements, so the addition of "medium" (micronutrient solution) led to a *reduction* of the SMA instead of the expected increase.

6.3.3 The Use of "Special Products" for Desulfurization

All biomass contains sulfur in different concentrations. The degradation of organic matter by the metabolic activity of the anaerobic microorganisms releases sulfide acid, H_2S, which is an inhibitor of the biological activity and corrosive for metallic parts. It is then necessary to desulfurize the biogas, mainly to protect the motor. The existing desulfurization technologies can be either biological—carried out by bacteria that precipitate the H_2S as free S—or chemical, neutralizing the acid with alkalis (lime, caustic soda, sodium carbonate, or bicarbonate) or ferric compounds (especially those containing iron in the trivalent state). Most European biogas plant constructors omit dedicated (i.e., external) desulfurizers, while in the USA, China, and India, it is the opposite case. The European biogas industry usually prefers the *in situ* desulfurization (i.e., in the digester itself). Said technique requires injecting a small quantity of air into the digester's headspace, necessary for the sulfur-reducing bacteria, but causing a slight reduction of the methane concentration in the biogas because of the added N_2, which represents 88% of the air injected. Such a method is economical (for the biogas plant constructor!) but not consistently fully effective in eliminating sulfur; hence in some cases, the plant manufacturer requires or specifies adding some "special products", generally the most expensive among the possible alternatives and for sure, always more expensive than their equivalent commodities available in the industrial chemistry market.

The cheapest and most environmentally compatible desulfurizer, which no manufacturer or "specialist" wants to recommend, is quicklime (calcium oxide, CaO), just because it is readily available on the market and allows the biogas plant owner to be independent. Quicklime reacts with the dilution water, forming spent lime (calcium hydroxide, $Ca(OH)_2$). This one can either react directly with the H_2S or with the CO_2 dissolved in the sludge, forming, respectively, chalk (calcium sulfate, $CaSO_4$) or calcium carbonate (limestone, $CaCO_3$). Calcium carbonate, in turn, reacts with H_2S and precipitates as chalk. The drawback of using lime as a desulfurizer is the sediment that can grow with time if the stirring of the digester is not vigorous enough. The presence of chalk and limestone does not alter the fertilizing properties of the digestate, since both compounds are usually present in soils.

Some companies promote the use of sodium bicarbonate as a desulfurizer. We already discussed the use of bicarbonate as a buffer in Section 6.2.1. Using it as a desulfurizer is analogous to using lime, with the difference that sodium does not precipitate. On the other hand, its presence in the digestate may reduce the fertilizing properties of the latter. The desulfurizers preferred by constructors and companies that manage biogas plants for third parties, and at the same time sell such products, are iron compounds. From a technical point of view, such a choice is correct because trivalent iron—present in ferric oxide (Fe_2O_3), ferric hydroxide ($Fe(OH)_3$), and ferric chloride ($FeCl_3$)—is very effective in selectively neutralizing H_2S. Iron compounds combine with sulfur, precipitating as insoluble salts, in the ways described by the following reactions:

1 Iron(III) oxide (ferric oxide)

$$3H_2S + Fe_2O_3\,(\text{in aqueous medium}) \rightarrow Fe_2S_3 + 3H_2O$$

2 Iron(III) hydroxide (ferric hydroxide)

$$2Fe(OH)_3 + H_2S \rightarrow 2Fe(OH)_2 + S + 2H_2O$$

3 Ferric chloride

$$3SH_2 + 2FeCl_3 \rightarrow Fe_2S_3 + 6HCl$$

4 Ferrous hydroxide (bivalent iron)

$$Fe(OH)_2 + H_2S \rightarrow FeS + 2H_2O$$

From a commercial point of view, the said technical choice generates significant revenues for companies selling "special desulfurizers" to biogas plant managers, who in good faith believe they are purchasing the best products to protect their motors from corrosion. Table 6.2 shows a comparison between three offers from different suppliers: an Italian constructor of biogas plants, a German laboratory service company that sells "special" products for biogas plants, and a wholesaler of standard commodities for the chemical industry. The first two have named their "special desulfurizers" with fantasy names, quoted as X and Y in this text for copyright reasons, but according to their technical safety sheets, both products are just iron(III) hydroxide. By the way, not of the best quality, as it contains other impurities.

The former example shows how scientifically sound information from the literature is being exploited to sell low-quality products at higher prices than the market average for the same commodity. It also demonstrates how the European biogas industry saves on installation costs by choosing the *in situ* biological desulfurization and not installing external desulfurization units, charging the operational cost of iron compounds to the biogas plant

TABLE 6.2

Comparison between Three Commercial Desulfurizer Products

Supplier	Product	Composition, According to Technical Sheet	Effective Fe^{+3} Content	Impurities	Price (Italy, Sept. 2016)
Italian constructor of biogas plants	Fantasy name X	$Fe(OH)_3$	36%–40%	Pb, Cd, Cr^{+6}, Cu, Ni, Hg, Tl, Zn	1.20 €/kg
Biogas-specialized German services company	Fantasy name Y	$Fe(OH)_3$	38%–41%	Pb, As, Cd, Ni, Cu, Cr, Zn	1.42 €/kg
Wholesaler of industrial chemical commodities	Ferric oxide (CAS N.1332-37-2)	Fe_2O_3	66%	Al_2O_3, SiO_2, MnO	0.87 €/kg

owner—and earning money on top—when the biological system is not able to bring the H_2S content in the biogas below the admissible limits.

6.3.4 Databases and Mathematical Models of BMP

Some European research groups and constructors of biogas plants promote large databases of BMP of many substrates as a fundamental tool for designing and managing biogas plants. There is a generalized belief, even in academic circles, that such databases are enough to predict the methane yield of any substrate under any condition, as if the BMP were a defined and invariable property of matter. The European Biogas Association (http://european-biogas.eu/) has collected the links to seven different Internet sites offering the possibility of free and paid access to such databases. Table 6.3 shows a comparison between the BMP of some usual biomasses published in three of these databases. The reader can draw their own conclusions about the usefulness of such data.

Should there be any doubt, the following experimental data will dissipate it: A sample of microcrystalline cellulose—a pure substrate with a well-defined theoretical BMP—was tested with an AMPTS II using two different inocula, sampled from two biogas plants running normally. The test was run in duplicate, with an error margin of 2%. Figure 6.7 shows the results: The difference between the methane yield of both inocula was 11%. Hint to solve the riddle: Inoculum A was sampled from a digester treating wastewater sludge, while inoculum B was sampled from an agricultural biogas plant, usually fed with bovine and swine manure and cereal silage.

6.3.4.1 Conclusion

If only the quality and biodiversity of the inoculum can induce 11% variation on the measured BMP, not considering any other experimental errors,

TABLE 6.3

Comparison between the BMPs of Some Common Agricultural Substrates for Anaerobic Digestion, According to Different Sources

	BMP According to Different Source [Nm^3/ton VS]		
Substrate	**KTBL (Germany)**	**LFL (Germany)**	**Biowatts (Germany)**
Corn silage	310–411	294–372	305–326
Triticale silage	281	259	259–606
Grass silage	318	295–329	315–330
Cattle manure	209	154–247	122–220
Pig manure	336–453	240	240–273
Poultry manure	293	325	275
Meat waste	583	—	733
Glycerol	751	425	425
Horse manure	—	165	165

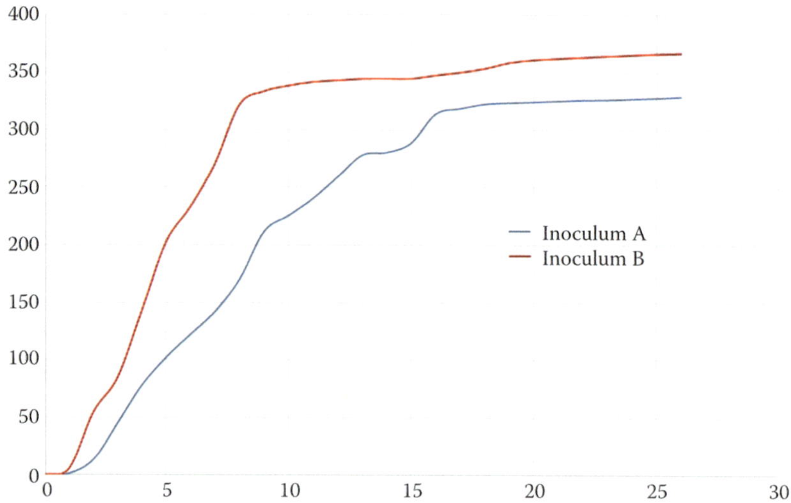

FIGURE 6.7
Anaerobic digestion of a single sample of pure microcrystalline cellulose with two different inoculums.

then: What is the reliability of tabulated values, not knowing the kind of inoculum, the measurement method, and the error margin of the test from where the said values come from?

6.3.5 The Conservation of the Inoculum

Norms on biological tests, such as the VDI 4630/2016 – Section 5.3, usually state that the inoculum must be sampled and conserved at 4°C until the moment of the test, to prevent the bacteria from dying and negatively affecting the test result. It is known to all biologists that bacteria and archaea are very resistant organisms. Some papers in the literature state that the inoculum for BMP is to be sampled and kept at room temperature before starting the test (e.g., Luna del Risco et al., 2011). Hence, the Cartesian doubt is entirely legitimate: Is it indeed necessary to keep the inoculum refrigerated until the start of the test? It is highly improbable that a biogas plant manager needs to conserve the inoculum for long periods, since he has the possibility of sampling it when necessary. The present paragraph was included in this book for the sake of researchers and also biological assistance professionals, who perform routine tests for commercial biogas plants. Both categories of laboratory operators may face the dilemma of sampling inoculum from the plant each time they need to carry out a test or keeping in the lab's fridge a certain quantity of inoculum. In both cases, they will have to bear a cost either for the displacement to the plant or for the space and energy required by the refrigerator(s). The author decided to perform a straightforward experiment

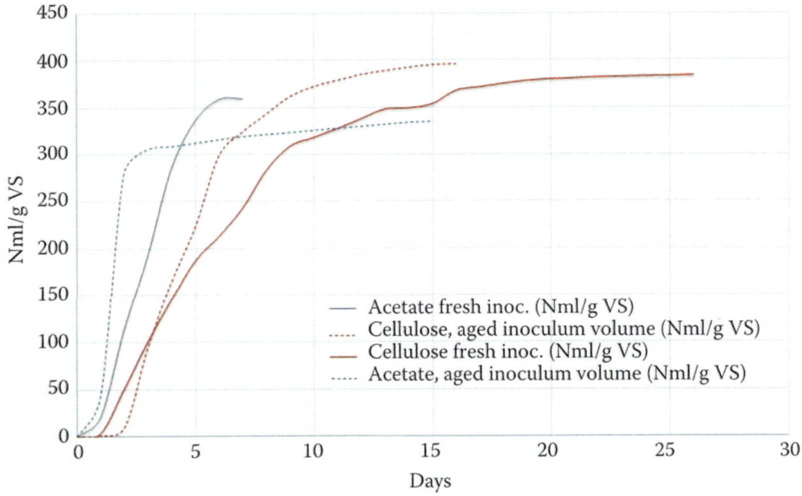

FIGURE 6.8
Hydrolytic and methanogenic activity of freshly sampled inoculum, compared to the same inoculum after 1 year of aging at variable outdoor temperatures.

to check if long periods of storage at uncontrolled temperatures can hamper the inoculum's bacterial activity. Two routine tests with cellulose and acetate were performed with fresh inoculum. Another sample of the same inoculum was kept in a hermetically closed plastic container, placed on a patio from November 2015 to November 2016, vented from time to time to avoid gas overpressure. The same routine tests with cellulose and acetate were then performed in December 2016. The results are shown in Figure 6.8.

The following can be observed:

1 The aged inoculum has a more extended lag phase. This is because after 1 year at varying temperatures, the microorganisms needed some extra time to adapt to the mesophilic conditions of the test.
2 After the lag phase, the aged inoculum showed a higher kinetic constant (steeper curves) than the one measured when it was fresh. This is because the microorganisms had depleted all the available substrate during the aging period and were starving.
3 The final BMP values of acetate and cellulose obtained with fresh and aged inoculum are not directly correlated to the age. One should have expected aged inoculum to yield lower BMP, which was indeed the case with acetate, but the opposite with cellulose. Anyway, the activity of the aged inoculum was still acceptable.

The extreme test described here shows that even keeping the inoculum during one full year, at temperatures varying from −2°C to 33°C (Northern

Italy), has not affected the hydrolytic and methanogenic activity too much. We can safely assume that keeping the inoculum indoors at room temperature for a few months will not compromise the results of the tests.

6.3.6 The "Equivalent Corn Silage Unit"

Most European biogas plants were designed to run on "energy crops", usually corn silage. Such a questionable design choice is the result of incentive policies adopted by the European Union based on a business model promoted by several key sectors of the German industry: the mechanical, electrical, chemical, and seeds industries, together with banks and investment groups. The reasoning supporting the use of corn silage as a feedstock for producing energy is an example of simplistic linear thinking: a single substrate—assumed as having a constant methane yield—a table with some standard parameters to control the process, a generator running 99% of its useful life at maximum power (unrealistic, but investors like it), hence return of investment fully warranted by State subsidies. A second factor that must not be overlooked is bureaucracy: the most sustainable use of anaerobic digestion is treating waste. Nevertheless, the paranoid European legislation about waste management and the ideological position of some political groups make it more difficult to obtain the necessary construction and operation permits. Consequently, using silage—which is an agricultural "product" or "byproduct" depending on the source—allows one to escape the additional bureaucratic complication of organic waste AD. The pressure of environmentalist groups, together with many studies showing that cultivating corn is not a sustainable way of producing energy, led to a re-evaluation of the subsiding policies. For instance, current (2025) regulations in Italy limit cereal silage ("agricultural products") to <10% of the total amount of "energy crops" in the biogas plant's feedstock; otherwise, the tariff paid for the energy falls from 233 €/MWh to 177 €/MWh. As a natural consequence, biogas plant managers began then sourcing other substrates that are labeled as "byproducts" and "cover crop silage". Since all their previous calculations were based on the yield of corn silage, this latter became a new measurement unit to calculate the digester's diet: the *equivalent corn silage unit* (*ECSU*). The *ECSU* is defined as the quotient between the methane yield of one ton of feedstock X and one ton of corn silage (assumed as a "universal constant", equal to 115 Nm^3/ton w.w.). The simple application of the syllogism method already demonstrates that the *ECSU* is not a reliable criterion:

1 The "standard" BMP of corn silage is 115 Nm^3/ton w.w. (FALSE! The BMP of corn silage is highly variable).

2 The BMP of a given biomass is X Nm^3/ton w.w. (may be true in a given moment, but strongly dependent on the moisture, which is variable in time).

3 Hence, each ton of feedstock X can replace X/115 tons of corn silage (may be true just for pure chance, but in principle this conclusion is FALSE).

The falsity of the first proposition was already demonstrated in Section 6.3.4 and with the example shown in Figure 6.7. The effective methane yield of a substrate in a given biogas plant at a given moment has nothing to do with the theoretical BMP; it is a combination of several factors, like the inoculum's bacterial biodiversity and the actual SRT of the digester.

The *ECSU* criterion may work if another silage is intended as a replacement for corn silage. Even if it is false that the BMP of corn silage is constant and equal to 115 Nm^3/ton w.w., the said value is quite frequent. Suppose the physicochemical features of "feedstock X" are similar to those of corn silage (e.g., similar density and moisture content). In that case, the *ECSU* criterion may work, even in plants having some biological issues. The reason is that the digestion conditions in the biogas plant are the same for corn silage and "feedstock X", then, the nondimensional ratio defined by the *ECSU* remains constant, independently of the effective methane yields of both substrates being lower than those obtained in the laboratory. Again, this is one of those cases in which the syllogism turns out to be true, but just by pure chance.

The following examples show the most common cases in which the concept of *ECSU* is unreliable for a correct formulation of the digester's diet.

6.3.6.1 Olive Mill Pomace

The BMP of this substrate varies not only as a function of the digester's bacterial ecosystem but also with the kind of olives, their degree of maturation, the oil extraction process, etc. Furthermore, olive pomace requires long digestion times of 45–60 days, while corn silage exhausts its methane potential within 15–20 days.

Hence, even if the "tabulated BMP" is nearly the same for both substrates (*ECSU* ≈ 1), replacing a given daily quantity of corn silage with the same amount of olive pomace can pose a serious risk of inhibiting the bacterial activity.

6.3.6.2 Fatty Waste and Glycerol

Such substrates have very high theoretical BMP; hence, *ECSU* >1, i.e., a small quantity of such substrates can replace large amounts of corn silage, an appealing possibility for biogas plant managers who need to buy biomass for their digesters. Nevertheless, the BMP of such substrates is a function of the OL (i.e., the I/S ratio), and experience shows that they can be potent inhibitors. Hence, replacing corn silage with its *ECSU* of glycerol or fatty waste will most probably inhibit the bacterial activity, eventually stopping

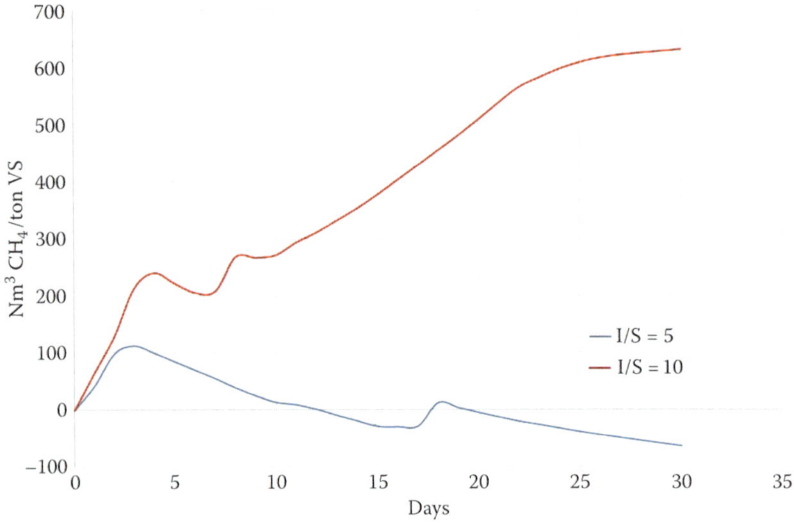

FIGURE 6.9
Inhibition caused by glycerol even at high I/S ratio.

or strongly reducing the energy production and its associated income. By pure chance, some biogas plants may not encounter any problem at all (e.g., digesters inoculated with sludge from sewage treatment plants, or plants fed regularly with fatty waste). From the author's experience, such plants are to be considered as exceptional cases, so the positive results observed in them cannot be generalized to all biogas plants. Figure 6.9 shows the digestion of the same sample of glycerol at different I/S ratios. The inoculum was sludge samples from a biogas plant, usually fed with corn silage.

6.3.6.3 Mixtures of Substrates

In the world of anaerobic digestion, 2 + 2 sometimes may be equal to 4, but in some cases, it can yield either 5 or just 3. The mixture of two or more substrates is the typical case when the total *BMP* is not proportional to the individual *BMPs* of the single components. Figure 6.10 shows the result of laboratory tests with mixtures of bovine and rabbit manure. The *BMP* of the mix depends strongly on the proportion of both substrates, resulting in a maximum when the VS of rabbit manure is four times that of cow manure (which in this test was stall manure, mixed with straw). Hence, should we try to apply the *ECSU* criterion to calculate a diet based on the mixture of said substrates, the theoretical result would be very different from reality. The reason for such an apparent paradox was already explained in Chapter 1: one of the parameters that influences the value of the *BMP* is the *C/N ratio*.

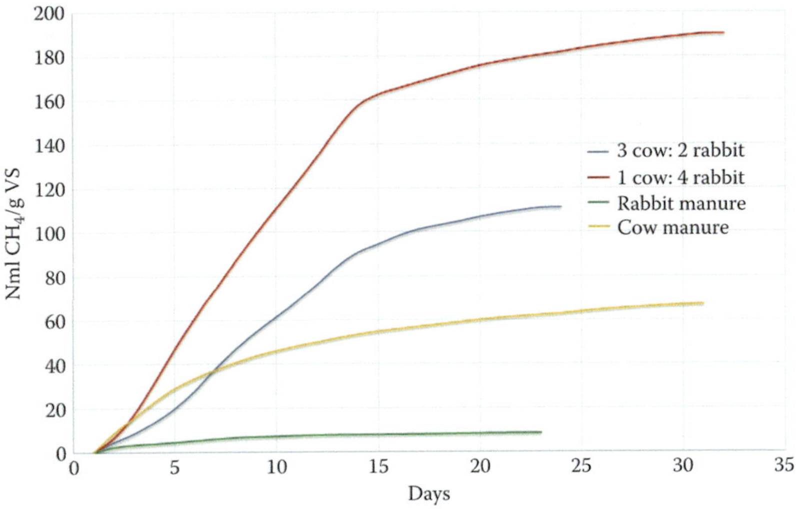

FIGURE 6.10
Mixtures with different proportions of the same substrates will yield different total BMPs.

6.3.6.4 Substrates with Low Concentration of Degradable Organic Matter

Another limitation to the use of the *ECSU* criterion for replacing silage with an "equivalent" substrate is the net digestion volume of the digester. For example, consider the AD of bovine slurry shown in Figure 6.11.

Such a substrate has 96% water, and its BMP is equal to 0.023 *ECSU* (i.e., 42.8 tons of slurry are needed to produce the same quantity of methane as 1 ton of corn silage).

A biogas plant consuming 50 tons/day of corn silage (approximately 1 MW power) would then require 2,141 tons/day of such slurry, which means occupying approximately 2,141 m^3. Observing Figure 6.11, we notice that the sample required 30 days for its complete degradation. Considering that the net digestion volume of average 1 MW_{el} plants (in Europe) is about 6,000 m^3, the impossibility of applying the *ECSU* criterion in this case is quite evident: the digester's HRT would be just 3 days, which is not enough to digest the slurry.

6.3.7 Estimating the Accuracy and Reliability of Data Published in the Literature

Quite often the biogas plant manager needs to include a new substrate in the diet of his digester, and faces the old dilemma: "What is the BMP of this feedstock?" We already explained that collections of BMP values are not reliable, so one could be tempted to look for answers in specific papers, where the authors and the test methods are known.

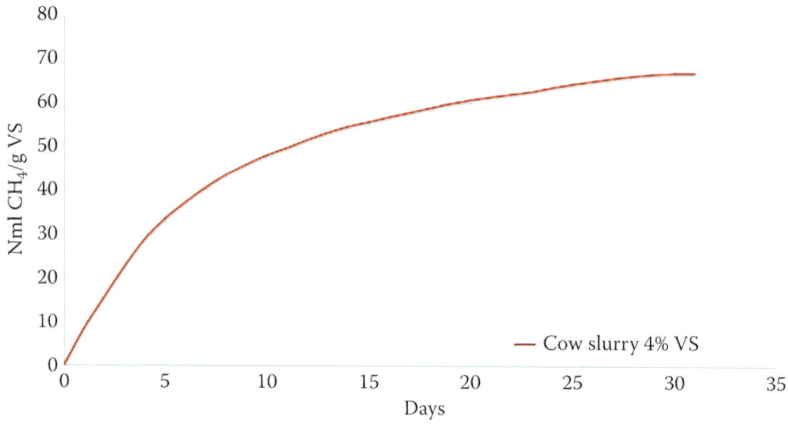

FIGURE 6.11
Anaerobic degradation of bovine slurry having 4% VS.

Despite the existence of specific norms[1] on assessing the uncertainty margin of laboratory measurements, a constant in all the biogas literature is the lack of such an analysis. Most of the papers on BMP of different substrates provide tables reporting the average values of several measurements and their dispersion. On one side, such values sometimes represent the "raw" gas production, which is unrealistic because it includes the moisture. From a metrological point of view, checking only the standard deviation of a set of measures is not entirely correct, since it is an indicator of the precision (i.e., repeatability) of the test, but not of its accuracy. The accuracy is paramount when the measured values must be employed for further elaboration—e.g., the energy value of the feedstock as a function of the gas production—since calculations amplify the error of the final result, according to the rules and theorems already explained in Chapter 2. Furthermore, gases vary their volume with temperature and pressure as well as their moisture. In many cases, the value presented is the total biogas production and the methane content, the latter being the result of a separate measurement that has its own error. Finally, the quality of a laboratory measurement can be determined only by comparing the uncertainty of the average value with the dispersion or standard deviation of the replicate measurements. If the error is of the same order of magnitude as the standard deviation, then the average value can be considered at the same time accurate and precise (the best possible quality of test with a given set of instruments). Should the error be bigger than the standard deviation, then the test can be considered precise (i.e., correctly performed) but not accurate (i.e., the average value is not reliable because the instruments and methods amplify the measure's uncertainty). In such a case, the lowest limit should be adopted as a conservative value for any calculation. Should the error be smaller than the standard deviation, then the average value is accurate and can be employed for calculations, even though the test is not precise (e.g., because the substrate is heterogeneous).

The following example illustrates how to analyze data from peer-reviewed literature to determine its reliability and usefulness for practical purposes.

The paper taken as an example is *Investigation of the methane potential of horse manure* by Mönch-Teder et al. (complete reference and link for free download in the Bibliography). According to the authors of the said study, horse manure has a very variable BMP for two reasons: the diet and metabolic state of the animal influence the composition of its dung; and the assortment of bedding materials employed by different stalls range from almost indigestible wood sawdust to fairly digestible wheat straw. The study is very detailed, as it tested and analyzed dung from many stalls and with different beddings. We will check one single value: horse manure from stall A containing straw, and then compare it with the remaining values from stalls B, C, D, and E, which contained straw too. The test protocol employed 30 g of inoculum and 600 mg of dried and pulverized manure, tested in triplicate, with only one blank. The BMP resulting from the normalization and average of the single samples from stall A was 0.176 Nm^3/kg VS ± 1.73%. Note that 1.73% is the dispersion coefficient of the three replicates (i.e., the standard deviation expressed in % of the average), not the error margin of the measurement. The procedure to estimate this latter is based on the following reasoning:

Graduated laboratory glassware can be labeled as class A (error = 0.5%) or class B (error = 1%), according to ISO 4788 (equivalent to DIN 12600). The paper states that the 100-mL syringes have 1-mL graduations, so we deduce they are class B and their absolute error is hence ±1 mL. Since they were filled with 30 mL of slurry and 600 mg substrate, the total volume of methane produced was:

$$V_{\text{methane}} = 0.6\ \text{g} \times 176\ \text{Nm/g VS} = 105.6\ \text{NmL}$$

Since the test temperature was 37°C, the volume of methane at this temperature was as follows:

$$V_{\text{methane}} = 105.6 \times \frac{310\ \text{K}}{273\ \text{K}} = 120\ \text{mL}$$

The paper does not state the composition of the biogas of each single measure. For simplicity, we can estimate that it contained 60% methane, so the total volume of biogas produced was in the order of 200 mL. Since the syringe only has 100 mL total capacity, and was filled with ~30 mL, the maximum admissible piston excursion is 70 mL. This means that at least one reading must be taken before reaching said point. If we observe the curves included in the paper, they were traced with 18 points. We can then deduce that each reading interval corresponded to an average of 11 mL of biogas production.

The total error of the sum of 18 individual measures can be then calculated with the rules of error propagation:

$$e_{\text{biogas}} = \frac{E_{v\text{biogas}}}{v_{\text{biogas}}} = \frac{18 \times 1\ \text{mL}}{200\ \text{mL}} = 9\%$$

Since the net normalized volume of methane, V_{methane}, results from a formula in which the ambient pressure, the temperature, and the percentage of methane are all factors, each one measured with a given error, then the total error of V_{methane} is as follows:

$$e_{Vm} = e_{\text{barometer}} + e_{\text{thermometer}} + e_{\text{gas analyzer}} + e_{\text{biogas}}$$

The error of the gas analyzer is not explicitly stated, but its brand and model are quoted in the paper. Checking the manufacturer's catalog, we find that the maximum error after a calibration is 2%. The researchers stated in the paper that they performed a calibration with 60% methane before taking each reading, so we can deduce that the 2% error in the gas composition measure was a constant throughout the test duration. The error of the barometer is not specified, nor is the barometer brand and model quoted, but we can assume that a reputed research center has laboratory-grade instruments. The absolute error of an electronic barometer for laboratory and meteorological use is about 20 hPa, so we can estimate that the relative error was roughly 2% throughout the duration of the test. The temperature accuracy of the thermostatic cabinet is stated as ±0.5°C, hence the relative error when measuring 37°C is 1%. The total uncertainty of the net normalized methane volume is then

$$e_{Vm} = 2\% + 1\% + 2\% + 9\% = 14\%$$

Observe that the paper provides no information about the gas production of the blank, which contributes to amplifying the error, as explained in Chapter 2. We will assume, as a simplification, that the researchers employed a preincubated inoculum, so that no error amplification needs to be considered. Furthermore, the paper does not provide information about the moisture content of the measured gas. At a temperature of 37°C, almost 13% of the gas volume in the syringe is occupied by water vapor. We will assume for simplicity that the researchers considered this systematic error and corrected it accordingly, since the paper states that a blank was employed. Anyway, given such high instrumental uncertainty, we can question the results published from a formal point of view: the authors of the paper should have stated their results according to ISO GUIDE 98-3:2008. The same defines the maximum number of decimal figures when expressing the results of calculations derived from measured values, which in this case is just two decimals:

$$V_{\text{methane}} = 0.176\ \text{Nm}^3/\text{kg VS} \pm 14\%$$

$$V_{\text{methane}} = 0.17 \pm 0.02\ \text{Nm}^3/\text{kg VS}$$

In other words, in this test it is impossible to discern if the Hohenheim instrument and ancillary equipment measured 0.15 or 0.19 Nm^3/kg VS, since any value within said range could be possible. Hence, expressing the average as 0.176 Nm^3/kg VS has no physical sense because the sum of the error margin of the measures contributing to calculating the said value is bigger than the millesimal order of magnitude. Such a misconception is a constant throughout the biogas literature: researchers seem to believe that expressing results with points and many decimals makes the said data more accurate.

6.3.7.1 Conclusions

1 This example shows that the data published in the paper are very precise (only 1.7% dispersion), showing high procedural skills and laboratory virtuosity of the researchers, but for practical effects the result is not accurate (14% uncertainty).

2 The BMP values of manure with straw bedding from the five different stalls are shown in Table 6.4. We can observe that the variation coefficient of the five averaged measures, 5.1%, is still smaller than the instrumental error. The Cartesian doubt is hence legitimate: Is such variability the result of the different fodder and metabolic state of the animals, as claimed by the researchers, or is it just a consequence of the high uncertainty of their measurement method? A sign that this suspicion may be true is contained in the same data provided by the

TABLE 6.4

BMP Values of Horse Manure with Straw Published by the Hohenheim Research Group and Further Analysis of the Author (Last Line)

Sample	BMP [Nm^3/kg VS]	Variation Coefficient [%]
A	0.176	1.73
B	0.173	0.68
C	0.163	5.97
D	0.191	3.45
E	0.175	2.98
Average	0.176	5.1

researchers. Looking at Table 6.4, we can observe that three samples gave the same result, 0.17 Nm^3/kg VS (we already demonstrated that the figures beyond the second decimal position are irrelevant), while the outliers, 0.19 and 0.16 Nm^3/kg VS, are both within the uncertainty margin of the test.

3 From a practical perspective, a biogas plant manager intending to feed his digester with horse manure from a stall with straw bedding may discard the outliers resulting from the paper and calculate his digester's ration with the representative average, 0.17 Nm^3/kg VS, or adopt a conservative attitude and employ the lowest value, i.e., 0.15 Nm^3/kg VS.

A final tip for those readers who eventually plan to employ horse manure containing straw bedding: Figure 6.12 shows a test performed by a biogas plant manager under the author's supervision. The instrument used was an AMPTS Light with stirred 2-l batch reactors. Being a test meant for practical purposes and not for research, a simplified protocol was adopted, consisting of a single sample and a blank. The inoculum was preincubated for just a few days; hence, the total uncertainty of the instrument is just 1%, but the amplification factor resulting from the inoculum's background activity was 4, so the overall uncertainty of this simple test is ±4%. It is not possible to analyze a standard deviation, because the test was carried out without replicates. However, the sample amount (120 g, no treatment at all) and the net volume of digestion (1,800 mL) are by far more representative of the real plant operation than the Hohenheim protocol. The resulting BMP was 0.174 ± 0.007 Nm^3/kg

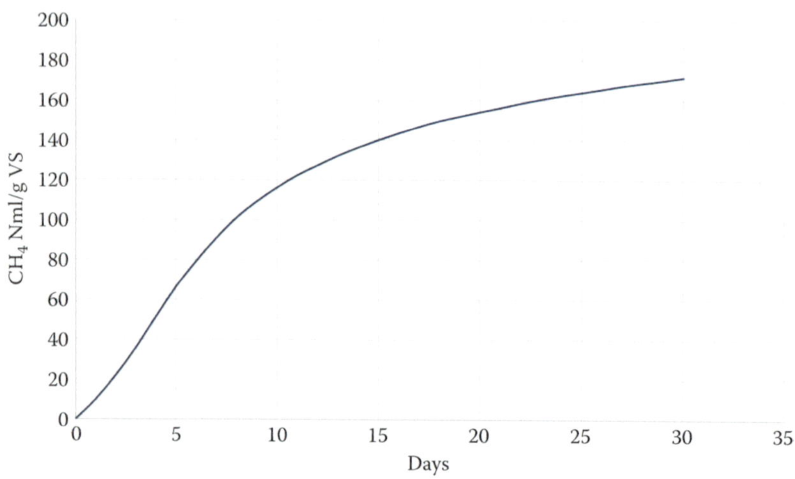

FIGURE 6.12
BMP_{30} of horse dung containing straw bedding performed by a biogas plant manager in his own laboratory, under the author's supervision.

VS. Coincidence or just correct application of the norms on measurement uncertainty? Note that, even being a test performed in an industrial environment by an unskilled operator, in this case, the third decimal position is relevant, since it is the same order of magnitude of the error.

6.4 Norms on the BMP Test Procedure for Industrial Biogas Plants

6.4.1 An Overview of the German Standard VDI 4630/2016

We already presented in Chapter 3, Table 3.2, a list of some norms on AD tests, pointing out that they are all focused on assessing the "ultimate anaerobic degradability" of wastewater, sludge, and urban waste. Therefore, their practical application to agricultural biogas plants is neither immediate nor pertinent, since the bacterial populations constituting the inoculum and the scope of the tests are different for both industrial sectors. In the waste treatment industry, the scope of AD is reducing as much as possible the pollutant OL and the volume of waste to handle, while the scope of all agricultural biogas plants is maximizing the energy production. When measuring the biodegradation through a respirometric method, each mol of gas corresponds to one mol of organic carbon, regardless of the gas composition. On the other hand, when the scope of AD is energy production, the focus of all tests is assessing the net methane production, all other gases being irrelevant. Germany has the merit of being the first country to publish a specific norm for the commercial biogas sector, with some focus on agricultural biogas plants running on energy crops. This norm, called VDI 4630, was first published in 2006 and reviewed in 2016. It is available in a bilingual German/English edition. The scope of the norm is to provide standard methods and criteria for conducting biological tests. Commercial biogas plants have gained economic importance worldwide since 2010. A significant challenge for plant managers has been the variability in BMP values measured by different laboratories, plant constructors, and researchers. The variability of the measured BMP values may sometimes lead to legal disputes when the plant underperforms, for instance, between plant owners and plant constructors or between plant owners and biomass vendors. The standardization of BMP tests was hence one of the challenges that the growing biogas industry had to face.

6.4.2 IWA's Protocol of BMP Test

The Anaerobic Digestion Experts Group of the International Water Association (IWA) published a set of recommendations on September 19, 2016. Such a proposal tends to overcome the worst problem all biogas industries and academic researchers face: the high dispersion of values resulting from BMP tests performed by different laboratories. The best outcome from several ring tests was a dispersion of 14% between laboratories.

6.4.3 The Italian Norm UNI/TS 11703:2017

The Italian standard is focused on BMP tests for assessing the energy potential of feedstock for AD. In broad lines:

1. The norm applies only to batch tests, wet digestion (5% maximum solids), and only to determine the BMP of solid waste and energy crops for energy purposes. Determining the residual greenhouse gas effect of digestate, analysing dry digestion processes, and tests with continuous reactors are all out of the norm's scope.
2. The protocol foresees a minimum experimental setup, one blank reactor and two sample reactors. The minimum reactor volume is 500 mL, so tests with syringes and "serum bottles" are excluded. Tests can be performed either with barometric or volumetric methods. The maximum acceptable error for barometric instruments is 1.5% of the end of scale.
3. Inoculum must be filtered with a mesh in the range of 1–5 mm in order to separate macroscopic solids and preincubated for at least 5 days or until reaching a given specific daily production.
4. Trace elements must be added if the laboratory is not 100% sure of the inoculum's "health".
5. Control tests with cellulose or acetate are not mandatory with each experiment, but should be performed periodically as an internal check of the laboratory's accuracy. They are recommended when the laboratory must test a biomass using an "unknown" inoculum, i.e., an inoculum whose actual methanogenic and hydrolytic activity has not been tested before.

6.4.4 Critical Analysis of the Flaws in the Existing Norms

All three cited norms and protocol contain logical flaws and recommendations based more on "we have always done it in this way" than on sound metrology principles and open-minded experimentation.

6.4.4.1 The VDI 4630

The first puzzling aspect when reading norm VDI 4630 is its extreme complexity. Reading between the lines, it appears evident that its redaction was conditioned by the two different schools of thought that have always influenced the concept of BMP itself. The first school of thought is the one we could call "theoretical", since it derives the BMP from chemical analysis and stoichiometric calculations. The second school of thought is the "experimental" one, since it calculates the BMP from biological tests. Paragraph 4 of VDI 4630, *Characterization of the substrate*, was clearly influenced by the theoretical school of thought, since it defines a series of preliminary chemical analyses that should be performed, with the scope of assessing the digestibility of the biomass. The said analytical procedures are, in general, complex, so it is implicit that they must be carried out in a laboratory adequately equipped. Hence, the possibility of self-management of the biogas plants is excluded, or at least strongly impaired, if the plant manager wants to comply with the norm fully. The limitations of such theoretical methods were already presented in Chapter 1, Section 1.4.3. In the author's opinion, including such useless tests as a norm prescription is an attempt to discourage the owner from self-managing the biogas plant, in favor of companies that offer biological management services, a cornerstone of the German biogas industry. It is spontaneous to ask oneself: What is the sense of performing a lot of preliminary analytical tests to check whether a feedstock is degradable, when it is more straightforward to load a reactor with a sample and check if it produces gas or not? For instance, consider the case of residual glycerol from the biodiesel industry. A chemical analysis will yield the result that it contains 95% biodegradable materials, but this does not imply that such a feedstock will produce 95% of glycerol's theoretical BMP in the AD plant. The tests performed by the author on crude glycerol from a biodiesel industry, already presented in Figure 6.9, show that the said substrate has a strong inhibiting power, at least for some inocula. The reasons are many: Quite often, the caustic soda employed for biodiesel production is neutralized with hydrochloric acid, forming common salt; furthermore, the first step in glycerol's anaerobic degradation is propionic acid. Both substances can be potent inhibitors of the anaerobic process if the inoculum is not robust enough.

The classification of glycerol as "fermentable with no limitations" in article 4.2.6 looks like an example of oversimplification bias, since it is not a universally applicable rule, as should be expected from a norm. Not all anaerobic plants are capable of digesting glycerol "without limitations", so laboratory tests are necessary to check the inoculum's suitability to avoid inhibition and loss of profit.

In the author's modest opinion, including in the norm a reference to the EU's Hygiene Regulation (EC 852/2004) and Animal By-Products Regulation (EC 1069/2009) and the legal classification of waste (article 4.6.8) is just a useless digression. EU regulations change with the political group in the government, while a technical norm is supposed to remain invariable over time because it is based on scientific evidence.

Another aspect, purely formal but of no less importance because of that, is the validity of the measurement units adopted as standard for the calculations and the expression of the results. The VDI 4630 adopts the liter as the unit of measure for gas volumes. From the formal point of view, this is not correct because the liter and its submultiples are not SI units (international metric system). There are norms on how to write norms, and the German experts who redacted the VDI 4630 seem to have ignored them. For instance, the *Bureau International des Poids et Mesures* very clearly states the correct definitions of the measurement units to express laboratory results, and the said rules are public domain:

> Cubic decimetre and litre
> The Comité International des Poids et Mesures recommends that the results of accurate measurements of volume be expressed in units of the International System and not in litres.
> https://www.bipm.org/en/committees/cg/cgpm/12-1964/resolution-6

From a practical point of view, the production of biogas—or biomethane—is measured in Nm^3 (m^3 in normal conditions, i.e., 0°C and 101.3 kPa). The liter cannot be considered a derived unit from primary units (m, kg, and s) and is deemed obsolete. Hence, the correct unit to define volumes within the text of a standard is the cubic meter (and its symbol, established by the BIPM, is m^3). Submultiples of the m^3 are acceptable. BMP values should hence be expressed in Nm^3/kg VS (primary SI units) or Nm^3/ton VS (acceptable derived SI units).

Chapter 5 of the VDI 4630 standard is dedicated to the methods for sampling the different kinds of substrates for industrial biogas production and the preparation of the samples before laboratory tests. The importance of the sample's representativeness for obtaining reliable test results is undeniable. Nevertheless, the contents of Chapter 5 seem somehow difficult to apply in practice. For instance, defining that sampling manure from a stall for a simple BMP test requires sending a specialized technician with a background in sampling methodologies—obtained in special training courses—appears as typical European overregulation. Furthermore, assuming that sampling manure is such a delicate and challenging task that deserves having a diploma, the "special courses" should be defined: Who can be the professors? What kind of school must provide such training? What is the duration and content of the courses?

The guidelines for the preparation of the sample prior to the test are reasonable: Coarse materials should be grounded to 10 mm or less. The following methodological steps—separating and weighing the inert materials, like small stones or sticks, sieving with a 10-mm mesh, breaking big lumps, etc.—are all scientifically correct, but put altogether they end up being too laborious and hence not practical. The content of article 5.4.3 sounds like pure technocratic fastidiousness: "To carry out size-reduction (crushing and grinding, for example) of the sample, a test screen with a 10 mm mesh (corresponding to DIN ISO 3310-1) and a hardwood cylinder with a diameter of roughly 10 to 50 mm are used. Size-reduction is effected by forcing the coarse material through the 10-mm screen". It is dubious that employing other trituration means, or a stick made of a different material or with a different diameter, could influence the result of a BMP test.

Chapter 6 of the VDI 4630 is a collection of quotations of other DIN and ISO standards and methods that are not standardized (e.g., the VFA/TA (FOS/TAC) test already discussed in Section 6.2.2.) or not directly applicable our helpful for the biogas plant (e.g., the Weender method, employed to calculate the available metabolic energy of fodder for ruminants, or the calorimetric method for measuring the lower heating value of biomass, which has little or no correlation with the BMP).

Considering that standards are not free and that the average cost of a DIN or ISO publication is about €200, a laboratory willing to perform BMP tests fully compliant with the VDI 4630 standard would spend at least €2.800 in norms to study, plus all the corresponding tools and instruments for each of the additional tests. Again, such overregulation is a penalty for small companies and tends to favor the already established (big) laboratories, both private and public.

Chapter 7 specifies how to perform biological tests and contains some contradictions. On one side, it states the use of airtight reactors, preferably made of glass, having volumes in the range of 0.5–2 L, built according to DIN 38414-8 or DIN EN ISO 11734 standards. Checking the said norms, it turns out that their scope is testing sewage water sludge, and the reactors specified do not have stirrers (they are either eudiometers or barometric reactors). While the lack of agitation is not a big problem when testing wastewater or sewage sludge, because the said substrates have a very low content of solids and thermal convection is usually enough to grant enough mixing, agricultural and municipal waste biogas plants work with high concentrations of solids, hence the stirring plays an essential role in the digestion of biomass. Ignoring the fact in laboratory tests makes it difficult to apply their results to a real-scale biogas plant; therefore, reactors with stirring mechanisms should be mandatory. Figure 6.13 is quite eloquent: a test was performed on microcrystalline cellulose. One reactor was stirred (60% power, 2 min on/4 min off), while the other was not. The error of the test was 1% because the inoculum's background production was negligible. The stirred reactors produced 6% more methane than the unstirred ones. The kinetic constants of both curves are pretty different, too, although the lag phase is the same for both.

At this point a trivial, yet legitimate doubt arises: What is the use of the VDI 4630 if it redirects to other (existing) norms that are meant for testing sewage sludge? In the author's personal opinion, the said test protocols are not useful for the management of agricultural or municipal waste biogas plants, because they do not reach the main scope of the VDI 4630 standard: "Defining a method that reduces the high variability of the results of BMP tests". The reason why the norm in question fails to reach its scope is that it tries to harmonize its content with other existing standards, losing specificity by admitting any typology of reactor and measuring instrument, regardless of its accuracy or suitability for the peculiar features of high-solids sludge. Such

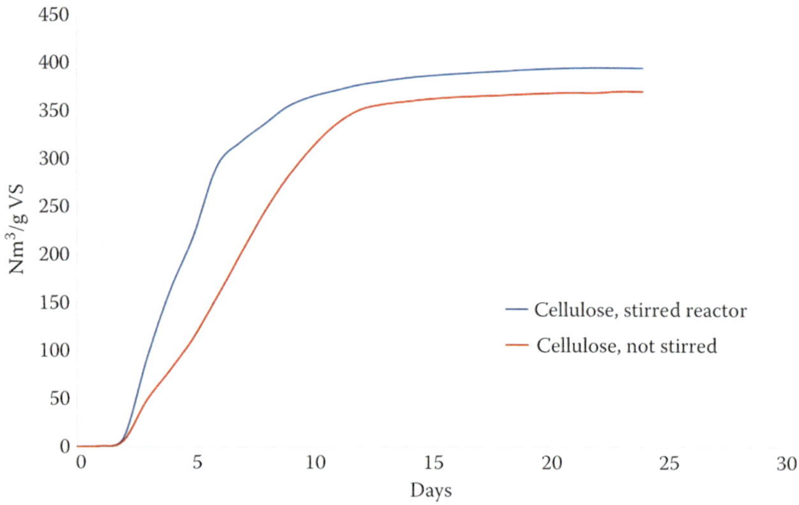

FIGURE 6.13
Influence of the stirring in the final BMP value.

a contradiction appears a few lines after the definition of the reactors, where the norm makes an exception to itself. It declares that the Hohenheim system, based on sets of 100-mL syringes, is acceptable. There is no scientific justification for such an exception, except for a self-referenced paper published by the Hohenheim University itself. We have already seen in Section 6.3.7 that the accuracy of measuring with such a system is very low. We also commented that Chapter 5 of the VDI 4630 is devoted to the method of sampling and ensuring the representativeness of the sample. It is difficult to figure out how measuring with a syringe that cannot be loaded with more than 1 g of lyophilized and pulverized sample can ensure the representativeness of the latter. Hence, including such an exception in a norm looks like a kind of protectionist concession to favor the "Made in Germany" brand.

Another critical aspect is the lack of compliance of Chapter 7 to the ISO/IEC Directive, Part 2—*Principles and rules for the structure and drafting of ISO and IEC documents*. This directive defines the method and protocol for redacting technical (ISO) norms and recommends that, in case two or more commercial methods are available for performing a given test or laboratory measure, the most accurate of them and its maximum admissible error margin must be defined. The error or uncertainty of measurement is fundamental in the case of the BMP assay, because the measured values have economic relevance, as explained with the following practical example. Suppose a biogas plant manager needs to determine the maximum price that a given biomass lot is worth. The maximum affordable price is a function of the net methane production under anaerobic conditions which, just for common sense, should be as close as possible to the real plant's operational conditions. According to

VDI 4630/2016, an eudiometer, a Hohenheim syringe carrousel, a barometric reactor with no stirrer, and a volumetric counter attached to a stirred reactor are all equally admissible instruments. Nevertheless, most of them measure in conditions that differ radically from those of a real-scale biogas plant, and each of them has a typical error margin, so a direct comparison of the test results is impossible. In other words, since the norm does not establish how to calculate the measurement error range, because it only considers the dispersion between replicates, each laboratory employing one of the test systems described earlier will obtain a single value as result, hence each laboratory can claim that its result is correct, even if it is wrong—or at least subject to a large margin of uncertainty. In any case, the VDI 4630 norm establishes that the results must be expressed in NL/kg VS, but does not consider the normalization error. According to research carried out at the University of Lund (Södeborg et al., 2014), when measuring the BMP with instruments lacking the feature of normalizing in real time, such as analog barometers, eudiometers, and Hohenheim syringes, in the most extreme cases, the normalization error can reach 80%.

Section 7.3.1 of the VDI 4630 norm provides very detailed formulas for calculating the BMP. These formulas include the correction of the moisture in the gas and the influence of the gas employed for flushing the headspace, which are a built-in feature of third-generation AMPTS. Such corrections are often overlooked in most of the biogas literature based on manual measurement methods and older electronic instruments lacking real-time normalization. The same section contains a method that must be applied when measuring the BMP of silage or any other biomass containing volatile matter (VFA, alcohol, etc). From the scientific point of view, such an approach is correct, but it adds complexity to the test because it is necessary to measure separately the solid and liquid volatile fractions, thus increasing the cost of the test.

The most disconcerting and metrologically wrong aspect of the VDI 4630/2016 is its Section 7.3.3: *Validity of the measures*. According to it, just performing triplicate tests and checking that the dispersion of the single values is <15% of the average (maximum 20% in the case of heterogeneous substrates, like municipal solid waste) is enough to ensure the reliability of the result. It is astonishing that in the country where the standardization of products and processes has its oldest tradition, a technical committee confuses two different concepts of metrology: precision (i.e., low dispersion of the measured values) and accuracy (i.e., the discrepancy between measured value and true value). Such confusion is even more difficult to understand if we consider that there are already specific international standards to define the precision and accuracy of a test or measure (see footnote at the beginning of Section 6.3.7). Furthermore, since including a positive control is not mandatory, situations may arise when the dispersion is within the standard's limits, while the digestion is incomplete because of the inoculum's low specificity to the substrate (see Section 3.5, Figure 3.1, and the example in Section 4.3.3). Section 7.3.2 of the VDI 4630:2106 provides indeed some qualitative criteria to check the shape of the digestion curve and decide whether a test may be unacceptable because of issues with the inoculum. Suppose that a test lacks a positive control of a nature similar to the sample and that the results have 12% dispersion and a curve presenting diauxia. In that case, the application of the dispersion criterion and the curve's shape analysis may lead to wrong conclusions. Should the test be discarded because of the diauxia or accepted because the dispersion is lower than the prescribed limit?

Chapter 9 of the VDI 4630 is dedicated to continuous tests. Although worthwhile for certain aspects of biogas plant management, such as quick restart strategies or transitioning between diets, continuous tests require significant labor over extended periods, making them more suitable for academic research than industrial applications. In the author's modest opinion, the preparation and performance of continuous tests should be the subject of a specific norm. In any case, to the present (2025) the VDI 4630 is the only standard that has the merit of providing guidelines on continuous digestion tests and an example (Annex H) of how to employ data from batch tests for the extrapolation to the real-size plant.

Annex H of the standard provides a helpful example of calculation; however, it also exemplifies academic presumptuousness and a lack of awareness of the existing ISO standards regarding the correct expression of test results. In Table H1, the daily gas productions are expressed to one decimal place. We must remember that, under the usual test conditions and employing the best instruments on the market, it is challenging to reduce the metrological test uncertainty below 3%, and the standard itself accepts an imprecision range of 15%. Hence, expressing a measurement like, for instance, 90.1 NmL/g VS has no physical sense, since the actual value could be any figure between 76 and 103, so the decimals are irrelevant. On the other side of the Atlantic, laboratory guidelines defined by the US NREL (National Renewable Energy Laboratory), although not having the status of a national standard, are unambiguous: *A final result should never contain any more significant digits than the least precise data used to calculate it*. Such guidelines are among the first results in any search on the Internet and can be downloaded for free (www.nrel.gov/docs/gen/fy08/42626.pdf). Amazingly, the experts who wrote the VDI 4630/2014 seem not to have considered these guidelines as valid, or even worse, they have ignored their existence.

6.4.4.2 The IWA Guideline for the BMP Assay

The document's introduction states that its scope is to overcome the high dispersion of BMP values encountered in interlaboratory tests (aka ring tests), a scope that, by 2016, the VDI 4630 had failed to achieve. In the first edition of this volume, the author had already predicted that the IWA protocol would fail too. The reason was factual truth already enunciated by Henry Ford almost one century ago: "If you always do what you've always done, you'll always get what you've always got". The 2016 IWA protocol perpetuated the same misconception as VDI 4630: confusing the instrumental error of the measurement with the dispersion of the replicates. According to the IWA proposal in 2016, the acceptance criterion of a test was the standard deviation of the replicates. For instance, the test described in our example in Section 6.3.7 would be acceptable according to the said criterion, while we have demonstrated that it has at least 14% uncertainty.

Another misconception is defining as acceptable any inoculum that reaches the prescribed minimum BMP of reference substrates like acetate and cellulose. We have seen in the real-life example presented in Chapter 4, Section 4.3, that reaching a minimum BMP with some reference substrate is a necessary but not sufficient condition for assessing the inoculum's quality. The suitability of an inoculum for digesting any unknown biomass is not only having the capability, but also being robust enough to carry out its task in the shortest time, i.e., the inoculum's *specificity* for that substrate, as will be explained later with an example.

The consequence of these flaws was that the ring tests conducted in 2016 and 2017 still yielded significant interlaboratory differences, and only 16 out of 62 results could be validated. In 2020, Hafner et al. defined new validation criteria. The commentary on these results, published in 2021, states that the application of the new criteria improved the comparability of results obtained with different instruments. Yet, significant differences between single laboratories still persist, which the authors attribute to internal protocols and peculiarities of the laboratories, pointing out that results obtained with manual methods, such as eudiometers, showed the largest dispersion. Couldn't it be that such differences are, at least partly, caused by the different inocula? On several occasions, the author has simultaneously tested samples using the same instruments and procedures with inocula from various sources, obtaining BMP differences as high as 20%. Nevertheless, all inocula satisfied the IWA's criteria for acceptability, since all were able to digest cellulose and acetate within the established ranges.

Another commonplace contained in IWA's protocol, and in all the biogas literature in general, is the fastidious method proposed for flushing the headspace of the reactors:

> During test preparation, flushing should be done with a mixture of N_2 and CO_2 that contains a similar share of CO_2 as expected in the produced biogas (e.g., 20%–40% CO_2; rest as N_2; v/v) to avoid a disturbance of the carbonate balance (Koch et al.). Flushing with pure N_2 should be carried out with care and only used for small head-space volume reactors.

Figure 6.14 shows a comparative test of cellulose digestion (with an AMPTS II, 500 mL reactors, hence "small headspace volume"). Three reactors were flushed with propane (a hydrocarbon more similar to methane than nitrogen) and the others were left unflushed. The difference is 3%, very close to the uncertainty margin of this test, which was 1%, so flushing or not flushing, or

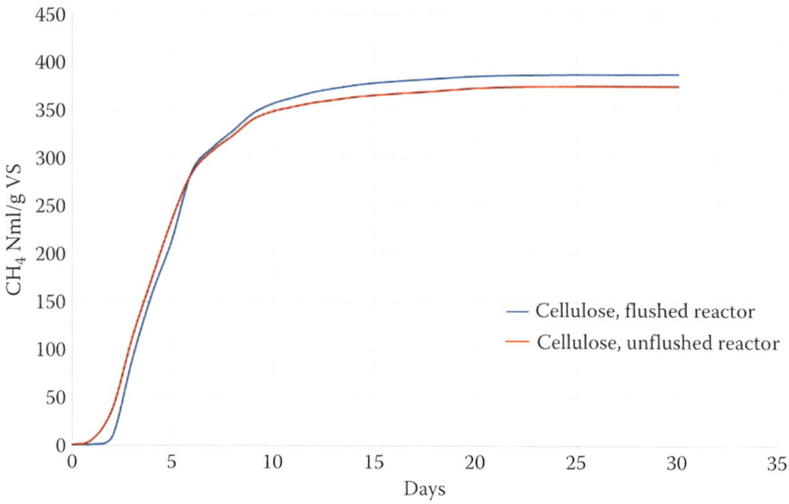

FIGURE 6.14
Influence of flushing the headspace volume.

flushing with gases not containing CO_2, induces differences in the results that are within the metrological uncertainty of the test. At least with volumetric methods and small headspaces. For sure, in tests performed with unstirred and analog reading apparatuses, such a difference is much smaller than all the other measurement errors put together. At least with "small headspace volume reactors" and cellulose as reference substrate, flushing seems to be practically irrelevant. By the way, the unflushed reactors show a shorter lag phase, which demonstrates that a bit of oxygen favors the hydrolysis of cellulose at the expense of producing some additional mmol of CO_2. The AMPTS II employed by the author to perform this test was sensitive enough to measure such small differences, but most other instruments in commerce would not be able to detect it.

The IWA protocol dedicates a paragraph to the inoculum-to-substrate ratio (ISR). It contains a generalization that cannot be considered an absolute rule, since it states,

> For easily degradable substrates where rapid accumulation of fermentation intermediates such as VFAs could lead to inhibition of anaerobic digestion, an ISR greater than or equal to four should be applied. For less degradable substrates, such as lignocellulosic organic matter, an ISR less than or equal to one can be applied.

We have seen that glycerol and fatty substances typically require high I/S ratios to ensure that all organic matter is adequately digested. Furthermore, VFA accumulation is usually an indication of either low acetogenic activity (in this case, propionic and butyric acids not being rapidly converted into acetic acid and cause inhibition) or low acetoclastic activity (in this case, the acetoclastic archaea are outnumbered or partially inhibited and are not able to degrade the acetic acid at the same rate it is being produced). Hence, just increasing the I/S ratio instead of ensuring that the inoculum is suitable for digesting a given substrate is a practice that perhaps can be acceptable for academicians—since they only look to get the final BMP—but is useless and even dangerous for biogas plant managers because it does not allow them to diagnose eventual bottlenecks in their specific plant's AD process. The suitability of the inoculum is connected to the following paragraph in the protocol: positive controls.

Positive controls are tests performed with reference substrates. The draft considers only cellulose and tributyrin (the main triglyceride composing butter). Such specification is too reductive, because the fact that an inoculum can degrade cellulose or tributyrin does not mean that it will be able to degrade any other substrate, and, once again, the time factor is not considered. It is easier to understand this concept with an analogy. Imagine you need to select the athletes who will participate in the Olympic Games. Imagine that, in order to simplify the qualification procedure, the selection criterion is defined as just being able to run 100 m. First of all, if the maximum time for running 100 m is not specified, anybody in any discipline can compete, even people not qualified as athletes. If the condition is set to run 100 m in 11 s, then a weight-lifting champion would probably be discarded, although being a champion in his specialty. Conversely, an athlete able to run 100 m in 11 s would qualify for competing as a swimmer or in weight lifting, even if not necessarily fit or trained for such disciplines. Assessing the suitability of an inoculum for BMP tests is similar to selecting an athlete for a pentathlon competition: the candidate must prove not only being able to perform different activities, but also perform *all of them* in acceptably short times, but not necessarily record-breaking times.

Figure 6.15 is an example of an inoculum that has evident biological issues, but still complies with the IWA guideline, as it reaches a BMP slightly higher than the minimum admissible (352 Ndm^3/kg VS). It is evident from the shape of the cumulated production curve that such an inoculum has problems because it reaches the plateau after 25 days, while cellulose is usually degraded between 12 and 20 days. Furthermore, the curve presents several humps and a sudden plateau, while a good inoculum would degrade cellulose, yielding a nearly perfect sigmoid (compare Figure 6.15 with Figure 6.14).

Figures 6.16 and 6.17 show, respectively, the digestion of starch and gelatin with the same inoculum. The reader can draw their own conclusions.

Another commonplace about inoculum is its conservation: the protocol states that it must be kept at operation temperature, but if this is not possible, and if the inoculum does not need preincubation, then it should be kept at least at room temperature, but for no more than 5 days. The experiment presented in Section 3.5 shows that such a disposition may be a bit excessive

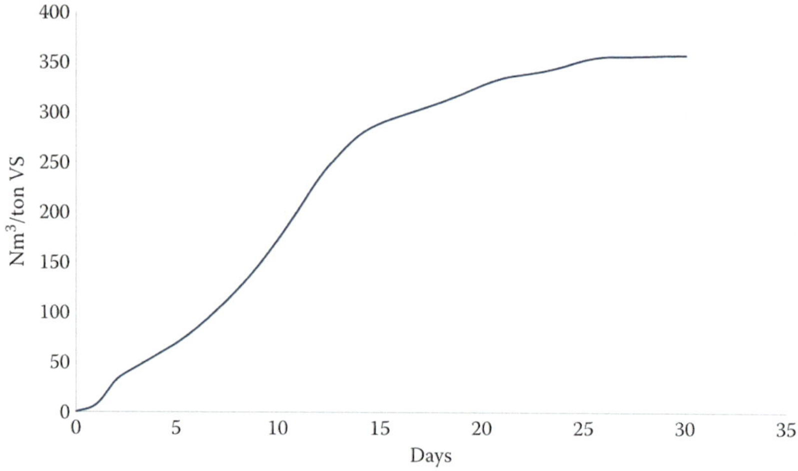

FIGURE 6.15
Anaerobic digestion of cellulose with a "weak" inoculum. Even if the final BMP reaches the minimum admissible value during the test with cellulose, such inoculum will probably yield underestimated BMP of other substrates.

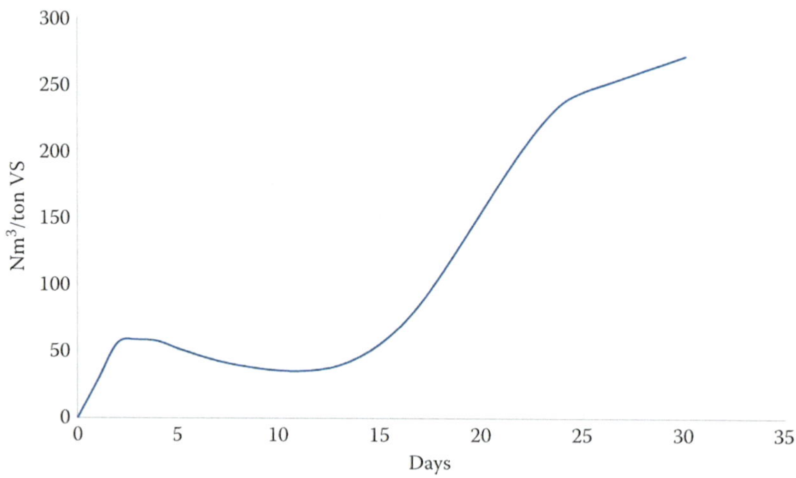

FIGURE 6.16
Digestion of starch with the same inoculum as Figure 6.15.

since longer times at room temperature can be admissible. The IWA protocol also contains a contradiction:

"The inoculum should, as much as possible, be used as sampled. However, some inocula may require preparation that can include sieving, dilution and pre-incubation".

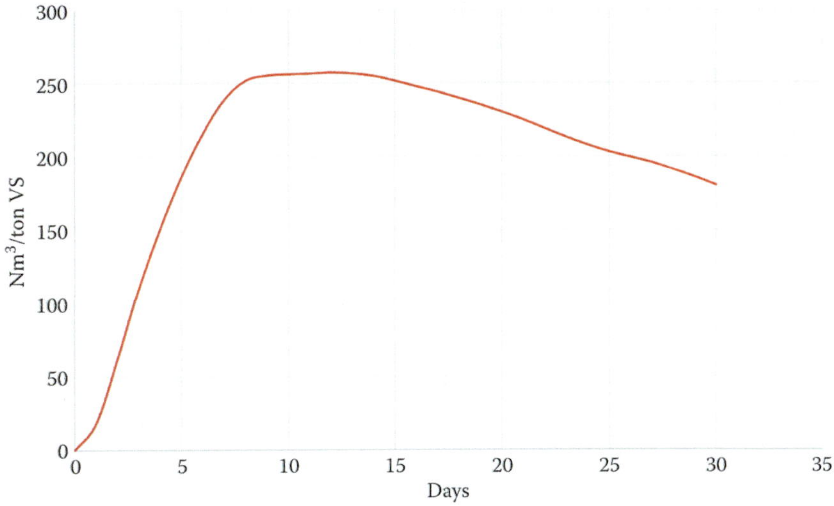

FIGURE 6.17
Digestion of proteins with the same inoculum as Figure 6.15.

So, what should we do? How do we know that the inoculum can be used "as it is" or when does it need preparation? This is also contradictory with one of the validation criteria stated in the same protocol: the endogenous production of the inoculum (blank reactor) should be less than 20% of the substrate's methane production (sample reactor).

Despite some flaws inherited from "the literature" and the VDI 4630, the IWA protocol has several merits: it is publicly available at no cost; it provides a set of rules that reduce the variability of results, and is open to future improvement. Anybody interested can collaborate via the official website of the DBFZ (Deutsches Biomasseforschungszentrum gemeinnützige GmbH), which has a section devoted to the BMP assay, www.dbfz.de/en/projects/bmp. The section covers all aspects, from the very basics of the assay to the refinements to adopt for each specific method and apparatus.

6.4.4.3 The Italian Standard UNI/TS 11703:2018

The Italian Standard UNI/TS 11703:2018 is the only one known to the author that includes a chapter on metrological error propagation, although the same is defined as "informative" (i.e., full compliance is not mandatory although in September 2016, the Italian UNI had adopted as a national standard the translation of the ISO GUIDE 98-3:2008 Guide to the expression of the measure uncertainty, with the name *UNI CEI 70098-3:2016* "Incertezza di misura—Parte 3: Guida all'espressione dell'incertezza di misura").

As in the VDI 4630 and the IWA guideline, the Italian standard considers neither the *specificity* of the inoculum nor *the time to digest the controls* as indicators of the inoculum's suitability for a given substrate; only its ability to reach the minimum reference value in tests with cellulose and acetate.

The conservation of the inoculum when not in use should be done at 4°C.

The proposed procedure to flush the headspace of the reactor is similar to the one proposed by the IWA, with further fastidiousness:

> The test and blank reactors must be flushed with inert gas, preferably a mixture N_2/CO_2 (50/50 or 80/20 in volume). N_2 can be employed as alternative. The flushing must be maintained during at least 5 minutes at a flow equal to 1 headspace volume per minute, i.e. the total volume of flush gas must be equal to 5 times the headspace volume, flowing during at least 5 minutes.

We already discussed the scarce influence of the flushing, at least in reactors with small headspace. Furthermore, one could argue: If biogas is composed of methane and CO_2, why should the flushing be done with "inert gas?"

6.5 Conclusions

"No book is so bad that it does not contain something good in it". The maxim, coined by Miguel de Cervantes Saavedra in his famous work *Don Quixote*, is as valid for novels as for technical literature. Our criticisms of the standards and the literature have the sole purpose of inspiring biogas plant managers and researchers to critically analyze all aspects of their BMP test results before making operational decisions or publishing a paper. The VDI 4630/2014 standard surely contains many positive aspects, for instance, a whole chapter on how to translate the results of batch tests to continuous industrial digesters, which no other norm includes. Nevertheless, in general terms, the practical application of VDI 4630 is too complex and expensive for industrial biogas plants, its content is too verbose, and rich in contradictions as well as conceptual metrology misconceptions. These latter make the results obtained with such a protocol quite questionable and partially contradictory to the ISO Guidelines 98 on the measurement of uncertainty and expression of the results.

The Italian standard on BMP has the merit of being more specific, focusing solely on the BMP assay for energy production from biomass, and its application is more straightforward than the German one; however, it still has several points that could be improved.

The IWA protocol still contains most of the logic flaws in the VDI 4630 and "the literature" in general. Since it is an "open" protocol, it may be improved in the future, provided that the issue of inoculum specificity is addressed and the error analysis is extended also to the metrological components.

Nine years have passed between the first edition (December 2016) and the redaction of the present second edition (July 2025). There is still much work to be done until a satisfactory standard for performing the BMP test is reached, which will finally eliminate disputes between the various stakeholders in the biogas industry and improve the quality of the scientific literature in general. The most challenging hurdle to overcome is the academic obstinacy, sometimes bordering on arrogance, as each research group presents its own internal procedures and instruments as "the correct way", arguing that such procedures and instruments lead to "results in line with those of the literature", but most of the literature on anaerobic digestion is obsolete and should be critically reviewed. At the time of Galileo, "the literature" said that the Earth was static at the center of the Universe. *Eppur si muove…* (Nevertheless, it moves), Galileo uttered.

From the author's practical experience, the only way to get reliable results from a BMP batch test is to comply with the following simple rules of thumb:

1. Volumetric methods with real-time normalization and mechanical stirring must be preferred, since they better mimic the real operation of a biogas plant. Stirring must be gentle, either continuous or on/off cycles.
2. Measure net methane directly (i.e., reactors connected to the measurement instrument through a caustic soda filter).
3. Reactors must have at least 500 mL in order to accept a quantity of sample that can be deemed "representative". This is particularly important for agricultural biomasses, which are usually heterogeneous. Reactors larger than 2 L provide no additional accuracy.
4. Use two blanks and triplicate sample reactors if these have less than 1 L capacity. For routine tests, one blank and two sample reactors are sufficient, provided that the reactors have at least 1 L of capacity.
5. Sieve with a 5-mm mesh and incubate the inoculum until its residual gas production is less than 0.03 Nm^3/kg VS of inoculum·day (or until the gas production on the nth day is less than 1% of the total cumulated volume up until then, both criteria are acceptable) to minimize the error amplification factor in the BMP formula.
6. The total cumulative error as well as the standard deviation of *both* blanks and samples must be checked, and outliers eliminated (if possible, repeat the test). A test is good (accurate *and* precise) if the total relative error and the variation coefficient are *both* less than 5%. A test is acceptable for industrial scopes if its relative metrological error is smaller than 5%, even if the variation coefficient is bigger (average value accurate, but not precise). A test is doubtful if its relative error is bigger than 5%, even if the variation coefficient is smaller than 5%.

7 Check the experimental setup to find small gas leaks before starting the experiment.
8 If the inoculum's microbial activity is not known in advance, then a complete set of positive controls must be performed in parallel with the BMP measurement of the sample biomass. Positive controls must be performed with acetate (always) and cellulose, starch, protein, or fat (depending on the type of substrate to be employed in the biogas plant) to verify that the relevant microbial groups are present and active. Positive controls can be performed in single reactors; there is no need for triplicates unless the test requires some legal value or we suspect a potential issue with the inoculum.
9 The quality of the inoculum is good if it passes a triple check:
   a The positive controls must reach at least 85% of the BMP of each reference substrate.
   b The shape of their cumulative volume curves must be sigmoid (short lag phase, no negative portions).
   c The indicative acceptable time to reach at least 80% of the minimum admissible BMP must be the following: 2 or 3 days for the test with acetate, 4–7 days for the test with starch, 10–15 days for the test with cellulose, 10–20 days for the test with protein, 20–45 days for the test with vegetable oils or fats.
10 If the positive control curves are not sigmoid, and/or the acetate takes more than 3 days for complete degradation, then trace elements must be added to the inoculum, it must be degassed again, and the test repeated.
11 Flushing with N_2 (or any other inert gas except pure CO_2), CH_4, biogas from the plant, or LPG is advisable but not a critical issue, at least for reactors with small headspace. A lighter's refill can is inexpensive and easy to handle, and propane is biologically inert, so this option is convenient despite not being cited in any standard. A volume of flush gas with at least twice the headspace volume should be enough.
12 pH is not a critical issue in agricultural biogas plants, since manure usually has more alkalinity than necessary.
13 Keep the same headspace and the same amount of inoculum in all reactors and blanks in order to eliminate the systematic error induced by the gas moisture or by the VS correction factor. If necessary, add distilled water to the blank. When it is not possible to keep the same amount of inoculum in all reactors, remember to apply the correction factor to the BMP formula.
14 Cumulated volume curves with two or even three plateaus (like the one discussed in Figure 4.3) are acceptable for mixtures of substrates. This is a normal feature of the batch test for compound substrates (sugars, protein, and fat), not a procedural error.

Note

1 ISO/IEC Guide 98-1:2024(en) Guide to the expression of uncertainty in measurement.

Bibliography

Abrol, I.P., Yadav, J.S.P., and Massoud, F.I., *Salt-Affected Soils and their Management*, FAO Soils Bulletin 39, Food and Agriculture Organization of the United Nations, Rome, 1988. www.fao.org/docrep/x5871e/x5871e00.htm#Contents.

Bakhov, Z.K., Korazbekova, K.U., and Lakhanova, K.M., The kinetics of methane production from co-digestion of cattle manure, *Pakistan Journal of Biological Sciences, 17(8)*, 1023–1029, 2014.

BIPM (Bureau International des Poids et Mesures), *Evaluation of Measurement Data—Guide to the Expression of Uncertainty in Measurement*, 1st edition, JCGM 100, Paris, France, 2008.

Bruni, E., Jensen, A.P., and Angelidaki, I., Comparative study of mechanical, hydrothermal, chemical and enzymatic treatments of digested biofibers to improve biogas production, *Bioresource Technology, 101*(22), 8713–8717. doi:10.1016/j.biortech.2010.06.108.

Burgstaler, J., Wiedow, D., Godlinski, F., and Kanswohl, N., Einsatz von Natriumhydrogencarbonat in landwirtschaftlichen Biogasanlagen, *Landbauforschung—VTI Agriculture and Forestry Research*, 4, 343–352, 2011. www.thuenen.de/media/publikationen/landbauforschung/Landbauforschung_Vol61_4.pdf.

Choong, Y.Y., Norli, I., Abdullah, A.Z., and Yhaya, M.F., Impacts of trace element supplementation on the performance of anaerobic digestion process: A critical review, *Bioresource Technology, 209*, 369–379, 2016.

Dupont, Biogas enzymes brochure. www.iff.com/grain-processing/carbohydrate-processing/saccharification-optimax/

FAO, Maize in human nutrition, FAO Food and Nutrition Series, No. 25, ISBN 92-5-103013-8, Rome, 1992. www.fao.org/docrep/t0395e/t0395e03.htm.

Garuti, M., Soldano, M., and Fabbri, C., Conducibilità elettrica, utile monitorarla nel digestato, *L'Informatore Agrario, 40*, 2014.

Gustavsson, J., *Cobalt and Nickel, Bioavailability for Biogas Formation*, Linköping Studies in Arts and Science No. 549, Water and Environmental Studies, Linköping, Sweden, 2012.

Hafner, S.D., Fruteau de Laclos, H., Koch, K., Holliger, C., Improving inter-laboratory reproducibility in measurement of biochemical methane potential (BMP). *Water*, 12, 1752, 2020. www.mdpi.com/2073-4441/12/6/1752 or https://doi.org/10.3390/w12061752

Helffrich, D., and Oechsner, H., The Hohenheim biogas yield test: Comparison of different laboratory techniques for the digestion of biomass, Landtechnik 2003.

Holliger, C., Alves, M., Andrade, D., Angelidaki, I., Astals, S., Baier, U., Bougrier, C., Buffière, P., Carballa, M., de Wilde, V., Ebertseder, F., Fernández, B., Ficara, E.,

Fotidis, I., Frigon, J.-C., de Laclos, H.F., Ghasimi, D.S.M., Hack, G., Hartel, M., Heerenklage, J., Horvath, I.S., Jenicek, P., Koch, K., Krautwald, J., Lizasoain, J., Liu, J., Mosberger, L., Nistor, M., Oechsner, H., Oliveira, J.V., Paterson, M., Pauss, A., Pommier, S., Porqueddu, I., Raposo, F., Ribeiro, T., Pfund, F.R., Strömberg, S., Torrijos, M., van Eekert, M., van Lier, J., Wedwitschka, H., and Wierinck, I., Towards a standardization of biomethane potential tests, *Water Science & Technology*, 2016. http://wst.iwaponline.com/content/ppiwawst/early/2016/09/19/wst.2016.336.full.pdf.

Holliger, C., Astals, S., de Laclos, H.F., Hafner, S.D., Koch, K., and Weinrich, S. Towards a standardization of biomethane potential tests: A commentary, *Water Science Technology*, *83*(*1*), 247–250, 2021. https://doi.org/10.2166/wst.2020.569.

ISO/IEC Guide 98-1:2009. Uncertainty of measurement—Part 1: Introduction to the expression of uncertainty in measurement. www.iso.org/standard/46383.html.

ISO/IEC Guide 98-3:2008. Uncertainty of measurement—Part 3: Guide to the expression of uncertainty in measurement (GUM:1995). www.iso.org/standard/50461.html.

KTBL-Heft 88, Gasausbeute in landwirtschaftlichen Biogasanlagen – Potenziale, Erträge, Einflussfaktoren, 2021. www.ktbl.de/shop/produktkatalog/11526.

Lebuhn, M., Liu, F., Heuwinkel, H., and Gronauer, A., Biogas production from mono-digestion of maize silage–long-term process stability and requirements, *Water Science & Technology*, *58*(*8*), 1645–1651, 2008.

Luna del Risco, M., Normak, A., and Orupõld, K., Biochemical methane potential of different organic wastes and energy crops from Estonia, *Agronomy Research*, *9*(*1–2*), 331–342, 2011.

Michener, B., Scarlata, C., and Hames, B., *Rounding and significant figures: Laboratory analytical procedure (LAP)*, issued July 17, 2005, Technical Report NREL/TP-510-42626, January 2008.

Mönch-Tegeder, M., Lemmer, A., Oechsner, H., and Jungbluth, T., Investigation of the methane potential of horse manure, *Agricultural Engineering International: CIGR Journal*, *15*(2), 161–172, 2013.

Petta, L., Giuliano, A., Porzio, V., and Gamberini, P., Con il pretrattamento enzimatico cresce la sostenibilità della filiera, *Terra e Vita*, *50*, 29–30, 2013.

Ploechl, M., Hilse, A., Heiermann, M., Quiñones, T.S., Budde, J., and Prochnow, A., Application of hydrolytic enzymes for improving biogas feedstock fluidity, *Agricultural Engineering International: CIGR Journal*, 11, 1-16, *IX*, 2009. https://cigrjournal.org/index.php/Ejounral/article/view/1529/1259

Södeborg, S., Nistor, M., and Liu, J., Towards eliminating systematic errors caused by the experimental conditions in biochemical methane potential (BMP) tests, *Waste Management*, *34*, 1939–1948, 2014.

Verein Deutscher Ingenieure, *VDI 4630-2014 Vergärung organischer Stoffe Substratcharakterisierung, Probenahme, Stoffdatenerhebung, Gärversuche (Entwurf—April 2014), ICS 13.030.30, 27.190.* www.vdi.de/en/home/vdi-standards/details/vdi-4630-fermentation-of-organic-materials-characterization-of-the-substrate-sampling-collection-of-material-data-fermentation-tests

7

Glossary of Terms and Abbreviations

AD: Anaerobic digestion A fermentative process in which microorganisms degrade organic matter, yielding methane and carbon dioxide as the ultimate products.

ADP: Anaerobic digestion plant

Alkalinity: A measurement unit of a solution's resistance to pH dropping. It is proportional to the concentration of bicarbonates or carbonates. When expressed in equivalent mg of calcium carbonate, it is called Total Alkalinity.

AMPTS: Automatic methane potential test system

Ash: The mineral fraction of organic matter.

Biological collapse: A condition in which one or more groups of microorganisms are inhibited, breaking the trophic chain of organic matter degradation. It features the accumulation of intermediate products (usually volatile fatty acids [VFA]) and limited or null methane production.

Blank(s): Reactor(s) containing only inoculum. The net production of methane is assumed to be the difference between the average production of the replicates and the average production of the blanks.

BMP: Biochemical methane potential Net quantity of methane produced by anaerobic digestion. When followed by a suffix, the latter indicates the time at which the BMP was measured; e.g., BMP_{30} means the BMP resulting from a 30-day test.

BOD5: Biological oxygen demand at 5 days It is a measure of the aerobic biodegradability of biomass. Sometimes, cited in the literature on wastewater treatment, but of little use for industrial biogas plants management, since there is no direct correlation between the aerobic and the anaerobic biodegradability or organic matter.

Buffer capacity: See alkalinity

COD: Chemical oxygen demand It is a measure of the quantity of carbon present in the organic matter. It can be employed instead of the volatile solids as a calculation parameter for the BMP test. See Chapter 8 for the equivalence between COD and VS.

CRT: Cellular retention time Similar to Solids Retention Time (SRT), but refers to the solids fraction containing live bacteria.

DOI: 10.1201/9781003676201-7

CSRT: Continuously stirred reactor tank (aka CSTR = continuously stirred tank reactor) The most diffused type of digester in the agricultural biogas industry.

Digestate: The remaining sludge after the AD process, mainly consisting of indigestible matter (lignin, minerals) and live microbial biomass.

DM: Dry matter The fraction of matter excluding water. Synonym of total solids (TS).

d.w.: Dry weight

E: Absolute error By definition, the difference between the real value of a magnitude and the value effectively measured with a given instrument or device.

e: Relative error By definition, the quotient between the absolute error and the measured value.

Error propagation: When calculating a magnitude using two or more individually measured magnitudes as input data for a formula, the individual errors tend to add, resulting in a calculated value that usually has bigger error (uncertainty) than the individual measures from which it originates.

FOS/TAC: Commercial name of the VFA/TA ratio test, from the German acronym *Flüchtige Organische Säuren/Total Anorganisches Carbonat*.

Headspace: Portion of the digester, or laboratory reactor, occupied by biogas.

HRT Hydraulic retention time: The average time that a volume of liquid biomass or dilution water remains in the digester.

Inoculum: Biomass composed of live microorganisms that can degrade organic matter under given conditions. Inocula can be aerobic or anaerobic, depending on the kind of process; e.g., yeast is the inoculum for fermenting either bread dough (aerobic) or beer (anaerobic). Ruminant's manure is both the inoculum and substrate for AD. Digestate is the usual inoculum for performing BMP assays.

LCFA: Long chain fatty acids Fatty acids have more than six carbon atoms. They are usually insoluble in water, tend to form emulsions, and tend to saponify. In general, LCFA are inhibitory for the AD process, and require a long time for their complete digestion (45–60 days).

Mix ratio: A measure of the intensity of stirring in a CSRT reactor.

NL or Nm^3: Liters or cubic meters in "normal conditions" Since the volume of gases varies with temperature and pressure, it is necessary to define at which conditions a gas volume has been measured. In the European biogas industry, the norms define gas volumes at 0°C and at an atmospheric pressure of 101.32 kPa.

ODM: Organic dry mass Synonym of VS, usually employed in the European technical literature (literal translation from the German Organische trocken Masse).

OL: Organic load = kg of VS (or COD) loaded to the digester/day

OLR: Organic loading rate = kg of VS (or COD)/m^3 of digester/day

ORP: Oxidation reduction potential (aka redox) A measure of the electrons exchange capacity between chemical species. The unit of measure is mV. If the ORP value is positive, the reactions are oxidative (in general aerobic); if negative, they are reductive (in general anaerobic).

pH Hydrogen potential: Nondimensional scale that measures the acidity or basicity of a solution. If pH = 7, the solution is neutral (distilled and degassed water). If pH < 7, the solution is acid (vinegar, beer, Coca Cola, lemon juice, demineralized water, hydrogen peroxide). If pH > 7, the solution is said to be basic or alkaline (caustic soda solution, tap water, sea water, ammonia, bleach, soap for dish washing machines).

Replicates: Each one of the reactors containing a given amount of sample biomass, whose average methane production is employed to calculate the BMP.

Short circuit (more correctly, hydraulic short circuit): Fluid dynamics condition causing an HRT much shorter than the nominal. It is equivalent to a virtual reduction of the digester's volume.

SL or Sm^3: Standard liters or cubic meters The concept is the same as that of NL or Nm^3, but defined as "standard conditions" at 20°C and 101.3 kPa. Not much diffused in the biogas industry, this is a standard from the natural gas industry.

Sludge: The digester's content, usually composed of live organic matter (bacteria and archaea, see inoculum) and partially degraded substrate.

SMA: Specific methanogenic activity Maximum amount of methane produced from sodium acetate (or acetic acid) per day, per g of VS of inoculum.

Substrate: The organic matter whose BMP is to be measured; feedstock of the biogas plant.

TOC: Total organic carbon Idem COD, but measured with a different method. Theoretically, TOC = COD × 12/32.

TS: Total solids contained in the substrate (both organic and inorganic).

Turnover time: A coefficient defined as the quotient between the volume of the digester and the recirculation flow. Employed as a control parameter in CSRT digesters using pumps as a stirring system, the desirable value is in the range of 20–30 min.

Unit gas flow: A coefficient defined as the quotient between the injected biogas flow/volume of the digester [in CSRT digesters stirred by blowing biogas from the bottom, desirable values in the range 0.24–0.3 (m^3/h)/m^3 of digester].

Unit power: A coefficient defined as the quotient between the power at the stirrer's shaft/volume of the digester (in CSRT digesters with mechanical stirrers, desirable values in the range 5–8 W/m^3 for perfect stirring).

V: Useful volume of the digester Portion of the digester effectively occupied by the active sludge.

VG: Velocity gradient By definition, VG = $[P_{stirrers}/(V_{digester}.\mu)]^{1/2}$ (where μ is the dynamic or absolute viscosity of the sludge). Theoretically, it is the best way to define the intensity of stirring, but it is practically impossible to measure μ, because sludge is a non-Newtonian fluid. Ideal values in the range VG = 50–80 s^{-1}.

VFA: Volatile fatty acids Short chain fatty acids (<6 C atoms). The most frequently found in normal digestion processes are acetic, propionic, and butyric acids.

VS: Volatile solids Fraction of the TS assumed as being completely digestible by the microorganisms. (VS = TS – ash)

w.w.: Wet weight VS can be expressed as a fraction of DM or as a fraction of w.w. Throughout this book, VS are expressed as a fraction of w.w., except when otherwise specified.

8

Helpful Tables for Quick Reference

This chapter presents a selection of tables previously introduced in the preceding chapters, along with some additional miscellaneous information, for the reader's quick reference.

8.1 Specific Weight of Silage as a Function of Its TS

The density of dry silage is fairly constant, independent of the ensiling conditions. The density of dry cereal silage is in the range of 235–245 kg/m^3, while that of dry grass silage is in the range of 250–260 kg/m^3. The density of silage increases with its moisture content, but not linearly, because moisture causes the fibers to swell and hence changes the volume. To calculate the density of wet silage, multiply the corresponding value of dry silage defined here by the corresponding coefficient shown in Table 8.9.

Example: Calculate the density of corn silage having 45% DM.

Moisture is equal to (100% – DM%), hence in this case it is equal to 65%. If you have not measured the density of dry silage, then you can assume 240 kg/m^3 as an average value (the density of dry cereals' silage). Multiply by 2.8, obtaining the wet silage's density as 672 kg/m^3.

8.2 Using Wine Vinegar to Carry Out the SMA Test

White wine vinegar usually contains about 6% acetic acid, so it is necessary to add a total of 4 g of caustic soda every 100 g of vinegar to neutralize it. After adding and perfectly dissolving 3 g, check the pH with the pH meter. Gradually add the remaining caustic soda, checking the pH at each step until it reaches at least 6.8. The resulting theoretical COD will be 6400 mg/0.1 L (i.e., 64,000 mg/L) and in the same way, the resulting theoretical VS will be 6 g/0.1 L, i.e., 6%.

Alcohol vinegar, employed as a scale and grease remover, may contain as much as 10% acetic acid. Please check the label for the exact acidity if you

DOI: 10.1201/9781003676201-8

TABLE 8.1

pH of Some Common Substances

Substance	pH
Hydrochloric acid, 1 M solution	0
Acid battery	1.5
Gastric juice	1.0–2.0
Lemon juice	2.4
Coca Cola	2.5
Vinegar	2.9
Antibacterial intimate soap	3.5
Orange juice	3.7
Beer	4.5
Acid rain	4.5–4.8
Coffee	5.0
Tea, healthy skin, intimate soap	5.5
Deionized water at 25°C	5.5–6.0
Oxygenated water	6.2
Milk	6.5–6.7
Distilled water at 25°C	7.0
Healthy human saliva	6.5–7.5
Blood	7.40–7.45
Water in a swimming pool	7.2–7.8
Sea water	7.7–8.3
Alkaline soap	9.0–10.0
Ammonia	11.5
Bleach	12.5
Lye	13.5
Sodium hydroxide, 1 M solution	14

TABLE 8.2

Content of C, N, and C/N Ratio of Some Common Biomasses

Substance	N (% d.w.)	C (% d.w.)	C/N
Corn straw	1.2	68	56.6
Cabbage	3.6	45	12.5
Hay	4	48	12
Alfalfa	2.5	40	16
Leguminous hay	1.6	40	25
Cow manure	1.6–1.8	30–40	17–25
Sheep manure	3.8	49–76	13–20
Stall manure (mixed with straw)	0.8	22	27.4
Horse manure	2.3	57.5	25
Solid swine manure	2.8–3.8	23–38	6.2–13.7
Swine slurry	0.4	4	10
Oat straw	0.5	40	80
Wheat straw	0.5	50	100
Barley straw	1	48	48
Layer hen's dung	3.7–6.3	31–35	5–9.6
Tomatoes	3.3	41	12.5
Kitchen waste	1.9	54	28.6
Corn stalks	1.4	44	31
Clover	3	39	13

TABLE 8.3

Equivalence Ratios between Volatile Solids (VS) and Chemical Oxygen Demand (COD) of a Few Pure Substances Usually Employed as Reference Substrates in Laboratory Tests

Substance	Chemical Formula	g COD/g VS (or Anhydrous Product)
Vegetal and bacterial biomasses	≈50% C	1.333
Glucose	$C_6H_{12}O_6$	1.067
Sucrose (saccharose)	$C_{12}H_{24}O_{12}$	1.067
Fructose	$C_6H_{12}O_6$	1.067
Starch	$(C_6H_{12}O_6)_n$	1.067
Cellulose	$(C_6H_{10}O_5)_n$	1.185
Proteins	Empirical formula $C_{400}H_{620}N_{100}O_{120}P_1S_1$	2.170
Acetic acid	CH_3COOH	1.067
White vinegar (6% acetic acid)	6% CH_3COOH + 94% H_2O	0.064
Sodium acetate	CH_3COONa	0.780
Propionic acid	$C_3H_6O_2$	2
Sodium propionate	$C_3H_5O_2Na$	1
Butyric acid	$C_4H_8O_2$	2
Sodium butyrate	$C_4H_7O_2Na$	1.6
Isopropanol	C_3H_8O	2.4
Vegetable oils (assumed as oleic acid)	$C_{18}H_{34}O_2$	2.89
Tributyrin	$C_{15}H_{26}O_6$	1.96

TABLE 8.4

Quantity of Daily Excreta and Approximate VS of Some Animal Species' Dung

	Daily Excreta (% of Live Weight)		Solids in Fresh Dung		Live Weight
Species	Manure	Urine	TS (%)	VS (%)	(kg)
Bovine	5	4–5	16	13	135–800
Buffalo	5	4–5	14	12	340–420
Swine	2	3	16	12	30–75
Rabbits	3	2	20	18	2–3
Sheep/goat	3	1–1.5	30	20	30–100
Hens	4–5		25	17	1.5–2
Humans	1	2	20	15	50–80

decide to use it instead of wine vinegar and adapt the dosage calculation as necessary. If adding vinegar directly to the inoculum, without neutralizing it with caustic soda, then calculate the total sample quantity to contain less than 3 g of acetic acid per liter of inoculum. Example: the reactor is filled with 900 g (approximately 900 mL) of inoculum, and you are employing alcohol vinegar with 10% acetic acid concentration. The maximum net quantity of acetic acid that you can add is then 3 g × 0.9 L = 2.7 g. The maximum quantity of alcohol vinegar you can add is then 2.7 g / 0.1 = 27 g ≈ 27 mL.

TABLE 8.5

Approximate Biochemical Methane Potential (BMP) of Some Common Agricultural Substrates (from Different Bibliographic Sources)

Substrate	Average BMP (Nm^3 CH_4/ton VS)
Swine manure	450
Bovine manure	250
Horse manure	460
Sheep manure	200
Stall manure (with straw)	225
Chicken dung	450
Corn straw	410
Corn silage	350–450
Rice straw	220
Grass	280–350
Vegetal waste	350
Sewage water	420

TABLE 8.6

Reference Substrates for Biological Activity Tests

Reference Substrate	BMP Range (min.-theor.)	Specific Bacterial Group
Cellulose	330–410 Ncm^3/g VS	Hydrolytic bacteria
Starch	330–410 Ncm^3/g VS	Hydrolytic bacteria
Glucose	330–373 Ncm^3/g VS	Acidogenic bacteria
Casein or gelatin	370–470 Ncm^3/g VS	Proteolytic bacteria
Propionic acid	330–350 Ncm^3/g COD	Acetogenic bacteria
Butyric acid	330–350 Ncm^3/g COD	Acetogenic bacteria
Long chain fatty acids (LCFA; vegetal oils)	700–1000 Ncm^3/g VS	Lipolytic bacteria
Acetic acid	330–350 Ncm^3/g COD	Acetoclastic archaea
Tributyrin	549–686 Ncm^3/g anhydrous (330–350 Ncm^3/g COD)	Lipolytic and acetogenic bacteria
Sodium acetate	330–373 Ncm^3/g VS (350 Ncm^3/g COD)	Acetoclastic archaea

N.B: The measured BMP is usually smaller than the theoretical one, because microorganisms convert part of the C present in the substrate into living biomass.

8.3 Monitoring the Electrical Conductivity of the Sludge

As it happens with monitoring oxidation reduction potential (ORP) and pH, this method is of little or no use, as demonstrated in Chapter 6, but thousands of biogas plant managers believe in it with an almost religious faith. For this reason, the author decided to include this table (Table 8.10).

TABLE 8.7

Typical Measurement Error Margins of Some Standard Instruments Diffused in the Market

Instrument	Max. Relative Error	Remarks
AMPTS II & III	±1% (+ e_{ga} if NaOH filters are not employed)	Normalization in real time, resolution 9 mL, can measure net methane or total biogas
AMPTS light II & III	±1% (+ e_{ga} if NaOH filters are not employed)	Normalization in real time, resolution 9 mL, can measure net methane or total biogas
Gas endeavour II & III	±1% (+ e_{ga} if NaOH filters are not employed)	Normalization in real time, resolution 2 or 9 mL, can measure net methane or total biogas
BPC Go	±1% + e_{ga}	Normalization in real time, resolution 9 or 2 mL, generic gas flow meter
MilliGascounter cell	±(3% + e_b + e_t + e_{ga} if NaOH filters are not employed)	6% = min. normalization error assuming constant *T* and *P* for the normalization formula
Eudiometer and similar water displacement instruments	±(e_v + e_b + e_t)	e_v = error of the graduated cylinder or burette, calculated as the quotient between the cylinder's error class and the reading
Oxytop bottle	±(1% + 1 hPa/P + e_T + e_v)	e_v = error in measuring the head volume of the bottle e_T = error of the temperature of the gas in the reactor's head space *P* = measured absolute pressure
Hohenheim syringes	±(5%–9% + e_b + e_t)	See an example of calculation in Chapter 6

e_b = error of the room barometer employed for the normalization of each reading; e_t = error of the room thermometer, idem; e_{ga} = error of the gas analyzer, including the uncertainty of the calibration mixture (when measuring total biogas production).

TABLE 8.8

Conversion factor between Nm^3 and Sm^3

1 Nm^3 = 0.947 Sm^3

8.4 Reference Values for Checking the AD Process with the VFA/TA (FOS/TAC) Test

As demonstrated in Chapter 6, Section 6.2.2, the said test provides coherent values only when the process is healthy, but it is absolutely unreliable when something is going wrong, since it measures all VFA *as if* they were pure

TABLE 8.9

Density Correction Coefficients as a Function of Silage's Moisture

Silage moisture (%)	40	45	50	55	60	65	70
Multiply by	1.6	1.8	2.0	2.2	2.5	2.8	3.3

TABLE 8.10

Alleged correlation between Electrical Conductivity and Methanogenic Activity, according to Garuti et al. Provided only for didactic purposes

Electrical Conductivity (mS/cm)	Methane Productivity (%)
<12	100
20	85
30	65
40	32
>50	0

TABLE 8.11

Reference Values of the FOS, TAC, and pH, According to Daniel and Baumgartner. Provided only for didactic purposes. The VDI 4630:20166 standard states that tables like this one are not universally applicable and depend on the digester type and feedstock.

	FOS [mg/L]	TAC [mg/L]	FOS/TAC	pH
Normal operation	2100	9000	0.23	7.8
Beginning of the acidification	3300	8300	0.40	7.6
Acidified reactor	4800	6800	0.71	7.0
Biological collapse	7300	3800	1.92	5.6

acetic acid. A false supposition cannot lead to a true conclusion, unless by pure chance (fundamental principle of logic). Again, since many people believe in such tests with an almost religious faith, here is a reference table. In the author's opinion, sometimes, it may be helpful to check the VFA (FOS) and the TA (TAC) values independently. An increase in TA may indicate, for instance, an accumulation of minerals or ammonia (this is particularly true in plants that recirculate a consistent amount of liquid digestate as dilution water). The German VDI 4630:2016 standard states that the optimum FOS/TAC quotient is plant-dependent, and within a plant, the optimum will be different for each digestion stage.

8.5 VFA Profile

TABLE 8.12

Normal, Maximum Acceptable, and Limit of Biological Collapse of the VFA Most Commonly Found in the Sludge, in mg/L (from Several Sources)

VFA	Normal	Max, Acceptable	Collapse
Acetic	100–200	2,000–2,400	4,000–6,000
Butyric	10–1,800	4,000–6,000	8,000–12,000
Propionic	5–15	900–1,000	>3,000
Total VFA	<2,000	2,000 (when the substrate is cellulose) 4,000 (when the substrate is sugar)	

TABLE 8.13

Minimum Concentrations of the Different Trace Elements Necessary to Maintain the AD Process's Stability According to the UNI/TS 11703:2018 standard

Element	Min. Concentration [g/m^3 to mg/L]
Boron (B)	0.003
Iron (Fe)	5.3
Nickel (Ni)	0.025
Zinc (Zn)	0.024
Cobalt (Co)	0.25
Copper (Cu)	0.18
Molybdenum (Mo)	0.004
Manganese (Mn)	0.14
Selenium (Se)	0.023

8.6 Oligoelements (a.k.a. Trace Elements or Micronutrients)

When the digester is fed with only cereal silage, the phytic acid contained in it chelates metals, making them unavailable for the archaea. It is then necessary to supplement the inoculum with such trace elements. Values in Table 8.13 have been taken from the UNI/TS 11703:2018 standard. It is highly advisable to keep the concentration of each mineral as close as possible to the minimum value, since they boost the activity of the acetoclastic archaea, but at the same time, the hydrogenotrophic archaea could be inhibited by their presence.

It is worth remarking that a chemical analysis showing the presence of trace elements with sufficient concentrations is not enough to grant the archaea's methanogenic activity. The reason is that the chemical analysis shows that a given element is present in a specific concentration, but it does not guarantee its bioavailability. Trace elements are all metals. They can either suffer chelation from phytic acid (or other eventual chelating agents) or react with the ubiquitous H_2S and precipitate as insoluble crystals. The safest action is performing a batch test in which a solution containing all the required trace elements will be added in different quantities to several test reactors, finding out which performs best (see Chapter 4, Section 4.4.3).

8.7 Guidelines for the Determination of VS

TABLE 8.14

Minimum Quantity of Biomass Sample, Necessary to Keep the Error of the VS Below 1%, as a Function of the Absolute Error of the Scale, E_{scale}, and the Moisture Content of the Substrate Under Test

Substrate Under Test	Approximate Moisture	Minimum Sample Quantity for $e_{(vs)} < 1\%$
Dry: straw, meals, oleaginous cake	Up to 40%	$400 \times _{scale}$
Semidry: silage, grass, urban waste, fresh manure	40%–80%	$1000 \times E_{scale}$
Humid: digestate, sludge, slurry	90%–95%	$3900 \times E_{scale}$

Index